AF361391

The Face of Immortality

The Face of Immortality

Physiognomy and Criticism

Davide Stimilli

State University of New York Press

Published by
State University of New York Press, Albany

Cover photo: Andrea Pisano, *Speranza,* from [Giovanni Paolo Lasinio], *Le tre porte
del Battistero di San Giovanni di Firenze,* Firenze 1821. University of Chicago Library.

The Northwestern University Research Grants Committee has provided partial
support for the publication of this book. We gratefully acknowledge this assistance.

For information, address the State University of New York Press,
90 State Street, Suite 700, Albany, NY 12207

Production by Diane Ganeles
Marketing by Susan Petrie

Library of Congress Cataloging-in-Publication Data

Stimilli, Davide, 1960–
    The face of immortality : physiognomy and criticism / Davide Stimilli.
        p. cm. — (SUNY series, Intersections)
    Includes bibliographical references and index.
    ISBN 0-7914-6263-3 (alk. paper)
        1. Face (Philosophy) I. Title. II. Intersections (Albany, N.Y.)

B105.F29S75 2004
128'.6—dc22   2003070443

10 9 8 7 6 5 4 3 2 1

To my parents

εὐτυχῶς μέν, ἀλλ᾽ ὅμως
τὰ τῶν τεκόντων ὄμμαθ᾽ ἥδιστον βλέπειν

A spirit pass'd before me: I beheld
The face of immortality unveil'd—
Deep sleep came down on every eye save mine—
And there it stood,—all formless—but divine.

Byron, Hebrew Melodies

Contents

Illustrations

Acknowledgments

I would like to thank the editors of the publications in which the following articles originally appeared, for having granted me permission to reprint them, in a revised form, as part of chapt. 2: "Character and Caricature," *Schede umanistiche* n.s., 1 (1996), pp. 51–72; chapt. 3: "Über Schamhaftigkeit. Ein Beitrag zur historischen Semantik einiger physiognomischen Begriffe," *Geschichten der Physiognomik: Text, Bild, Wissen*, eds. Rüdiger Campe and Manfred Schneider (Rombach: Freiburg i. Br. 1996), pp. 99–123; chapt. 4: "The Circumambient Air: or, Several Attitudes About Breathing," *Atenea* 18 (1998), pp. 131–137; chapt. 5: "Daimon and Nemesis," *RES: Anthropology and Aesthetics* 44 (2003), pp. 99–112, © The President and Fellows of Harvard College.

I would especially like to thank Rodolphe Gasché for welcoming my book in his series.

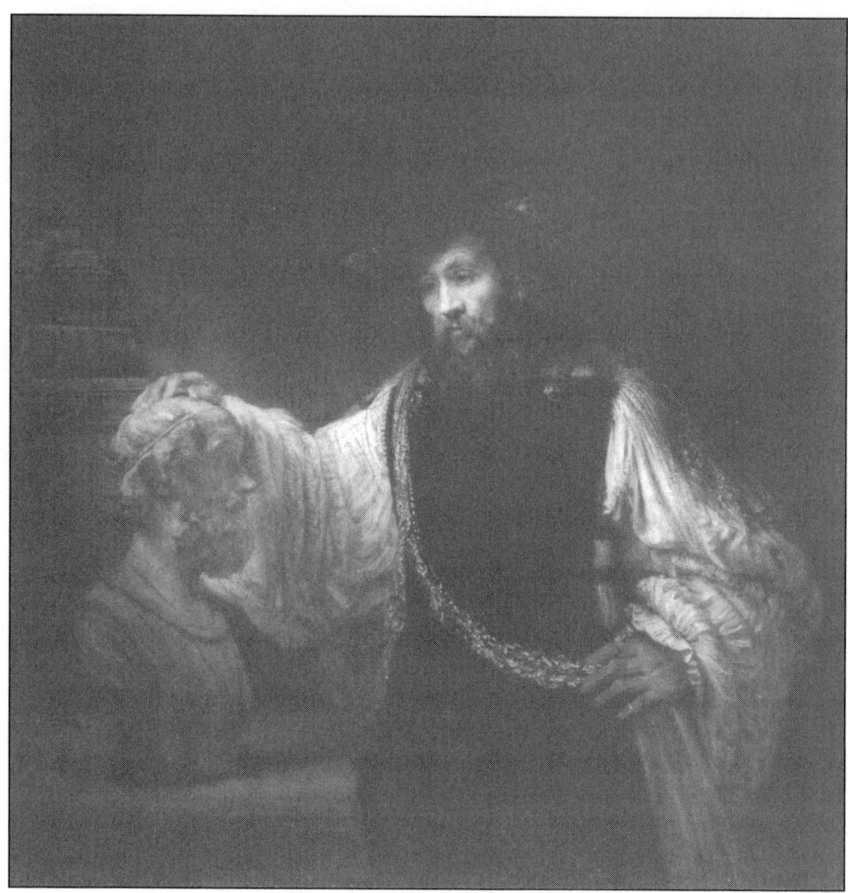

Fig. 1. Rembrandt, Aristotle with a Bust of Homer. All rights reserved. The Metropolitan Museum of Art, New York. Purchase, special contributions and funds given or bequeathed by friends of the Museum, 1961. (61.198)

Introduction:
The Strategy of Immortality

Of Immortality
His Strategy
Was Physiognomy.

Emily Dickinson

Socrates shies away from passing an aesthetic judgment over Charmides when he first meets the youth who gives the title to an early Platonic dialogue. He justifies his caution by warning that he is "no measurer," when it comes to beautiful people, but rather the proverbial "white line," which is, of course, useless when marking off measurements on *white* stone or marble; for, Socrates confesses with his customary irony, "almost everyone who has just grown up appears beautiful to me." He agrees, though, with all the bystanders that Charmides has "a fine face (literally, is well-faced: *eyprosōpos*)," but they are not satisfied. Everybody keeps staring at the youth as if he were a statue. His body is certainly more alluring to them than his face: "if he would consent to strip," one says, "you would think he had no face" (literally, he were faceless: *aprosōpos*), "he has such perfect beauty of form (*eidos*)." If exposed, Charmides's body would efface his face in the eyes of the viewers.[1]

This anecdote, I contend, is more than just the tale of an occasional infatuation. It accounts for the prosopagnosia, the face-blindness to which Western culture seems to be liable.[2] A pre-eminence of the figure over the face is undoubtedly the legacy of Greek humanism. Hans Castorp, the protagonist of Thomas Mann's *Enchanted Mountain*, echoes Settembrini, the Italian humanist, one of his two mentors in the novel, when he argues that "the Greek sculptors did not worry much about the head, what mattered to them was the body, that was perhaps what humanistic meant."[3] Castorp is avowedly a *dilettante*; his musings, though, would not have sounded naive

1

even if put in the mouth of an art historian: an authority such as Bernard Berenson could straightforwardly declare that

> so unnecessary do I find facial expression, and indeed, at times, so disturbing, that if a great statue happens to be without a head, I seldom miss it; for the forms and the action, if both be adequate, are expressive enough to enable me to complete the figure in the sense that they indicate; while there is always a chance that the head, in works of even the best masters, will be overexpressive.[4]

"Overexpressive" might hardly strike our contemporary taste as a criticism. Yet, in spite of the seeming casualness of their remarks, both Mann's character and Berenson were restating, almost word for word, one of the fundamental tenets of the grand style, as Sir Joshua Reynolds had codified it in his *Discourses on Art*, the manifesto of classicist aesthetics. Reynolds writes in the X Discourse:

> As the general figure presents itself in a more conspicuous manner than the features, it is there we must principally look for expression or character; *patuit in corpore vultus* [. . .] The face bears so very inconsiderable a proportion to the effect of the whole figure, that the ancient Sculptors neglected to animate the features, even with the general expression of the passions.[5]

The price humanism has to pay in order to establish the dignity of the human figure,—to the point that the gods themselves wish to look human[6]—is the effacement of the face, the banishment from its features of even "the general expression of the passions."

In arraying his Latin source, however, Reynolds mispells it: Statius wrote *latuit in corpore vultus*, which Reynolds's contemporary Joseph Spence freely amplifies as follows: "the whole Beauty of his Shape [. . .] exstinguished the Beauties they had before so much admired in his Face."[7] More literally we may render: the face hid in the body. I take the careless spelling in Reynolds's quote, the disregard for the letter that he so betrays, be it intentional or just a *lapsus calami*, as a symptom of his utter disregard for the face: it is obviously irrelevant to Reynolds whether the face is latent or patent in the body; what matters to him is that, either way, the body overshadows or outshines the face.

A face is no body, *personne*. Ominously, the same adjective Plato uses apropos of Charmides, *aprosōpos*, was later used in Greek law in reference to slaves, those who have no face, hence no legal person.[8] I take physiognomy to name the resistance to such an obliteration of the human face. In the pages that follow, however, I am not advocating the legitimacy of what Kant labelled "the art of spying the inside of man,"[9] nor indulging the "physiogno-mical QUIXOTISM" a B-novel of the early ninenteenth century diagnosed as "MORBUS INSANABILIS."[10] While conjuring up its name, I wish to elicit a different understanding of physiognomy and to advocate another physiog-nomy than that complicit with the very tradition of obliteration I am denounc-ing. Hence the usage of the term "physiognomy" in the context of my discussion entails an ambiguity of which the reader ought to be mindful.[11]

Throughout this book, I am concerned with the language we use to talk about the face more than with the language of the face *per se*, and I am more interested in the historicity of language than in the natural and/or social history of the face.[12] A second anecdote from the *Charmides* has been very often quoted in the literature on physiognomy, especially since Addison's essay in the *Spectator* (1711) made it current in the European-wide debate leading up to Lavater's ephemeral renown. Socrates proceeds to question Charmides in order to test whether his undeniable beauty of appearance corresponds to an interior beauty, which to him, as we may expect, is far more important. Socrates starts by inviting the youth to simply speak: "speak, that I may see thee."[13] This imperative is quoted again and again by the critics intent on dismissing the interpretive claims of physiognomy: man truly re-veals itself through language, not through the face.[14] Language is the true face of man, for language is the face of the soul, and not just of the body: *oratio vultus animi*, a sentence Leo Spitzer elected to sum up his credo as a critic.[15] Unfortunately, Spitzer misquotes his source, as well: Seneca meant *oratio* to be the *cultus* of the soul, namely, and not its *vultus*. Spitzer's mispelling is a sobering reminder that the physiognomy of language is not necessarily more transparent than the language of physiognomy. We mispell words as easily as we mistake faces. Werner Kraft more persuasively justifies the Socratic imperative when he writes apropos of Kafka that

> the essence of man is manifest in the face (*Gesicht*) and hidden in
> language; but since every manifestation for man is mere appearance,
> he can only be known in an essential manner in language.[16]

In language, though, the essence of man is latent or, at least, as little patent as in the face. Certainly, no immediate access to such an essence is to be gained through either face or language.

A face is a vision. This premise is almost obvious in German, in which the word *Gesicht* has both meanings, or in ancient Italian, in which *viso* (< Lat. *visum/visus*) is both the faculty and the object of vision.[17] Yet, when Rilke writes in the opening pages of *The Notebooks of Malte Laurids Brigge*: "A face is a face (*Gesicht ist Gesicht*)," his is not a tautology. It means, as he writes shortly before: "I am learning to see" (*Ich lerne sehen*).[18] In turn, what he implies is that a vision is not yet a face. A vision becomes a face only through language. Dante cannot translate (for him, as well as for the reader) his vision of Beatrice into a face, his "viso" into her "viso," because, even if he sees her truly *vis-à-vis*, her beauty "transfigures itself" (*si trasmoda*) and thereby evades the figurative power of language.[19] Within our mundane sphere, however, a face is always a prosopopeia, the imposition, brought about by language, of a face to a vision. Aristotle hints at such a process in the opening page of the *Physics*, where he suggests that the acquisition of language necessarily blurs in the eyes of the children the outlines of even those faces they most dearly love, and they end up "by calling every man 'father' and every woman 'mother.' "[20] This case of early prosopagnosia suggests that language per se is not the remedy to our face-blindness. Instead of making them more visible, language effaces faces by imposing a persona on them.

Physiognomy, I suggest, may point the way out of the impasse between the prosopagnosia of vision and the prosopopeia of language. Henry More, the Cambridge Platonist, renamed it "prosopolepsia," by transliterating a New Testament term the Vulgata renders as *acceptio personarum*,[21] and Tyndale "parciality." The Hebrew verb נָשָׂא פָּנִים, after which the Greek noun was probably coined, refers to the gesture of lifting up the forehead of somebody kneeling in front of us.[22] But to More, as the translation by Tyndale also implies, the word had a negative connotation: he uses it to refer to a minor vice in his system of ethics, the inclination to pass a judgment over somebody just at first sight. As I take it, physiognomy is unabashedly the name of such a parciality toward the face, without any negative connotation attached: the acknowledgment of the uplifted face, its recognition as human at first sight. In spite of its recurrent claims to the status of a science, physiognomy is indeed bound to remain a prosopolepsia, an acceptance, or just a reconnaissance, of the other's face *prima facie*: we do not reach any knowledge through physiognomy, we can only acknowledge faces, or recognize them. Recognition is "that which is sweetest when we meet face to face," Seneca writes in one of his letters,[23] but no knowledge is at stake in such an encounter: the relationship to the face of the other, as Emmanuel Levinas has persuasively argued, is never reduceable to a mere relationship of knowledge.[24]

Homer has a word for the sense that allows a mortal to recognize a divine countenance in disguise. That word is *noos*, which is used in reference to this physiognomical capability in the Homeric poems, before becoming the common noun for understanding in later Greek.[25] As applied to the human countenance, physiognomy is then a secularization of the ability to recognize the gods, but is also the dawn of understanding as such. We can then understand ourselves, I hope, how Euripides could call recognition "a god" in a verse of his play *Helen* that has been a *crux* to the interpreters, precisely because of their failure to see in the Aristotelian *anagnōrisis* anything more than a theatrical device. The protagonist invokes the gods to witness as she deifies recognition itself: "You gods! For recognition *is* a god."[26] Here the invocation is certainly not meant to invite the appearance of a *deus ex machina*, but rather to remind us that the recognition of a human face is always a divination, the possible recognition of a divine in a human countenance. And in recognizing as such we are ourselves recognized as god-like, for every face might be a god's. "How could we see the light, if the eye were not sun-like?"[27]

The German scientist Wilhelm Ostwald mocked Goethe's (revival of Plotinus's) rhetorical question by suggesting that, if we apply the same principle to reading, it becomes patently absurd: in order to read, the eye would have then to be ink-like.[28] But the paradox is only apparent. When confronted with particular obscure handwritings, the Renaissance philologists resorted to the principle "it is necessary to divine, rather than to read (*divinare oportet, non legere*)," which was misunderstood as if it were a loose principle of interpretation,[29] but the translation "to guess" would be almost blasphemous here. "To divine" is the proper term when we take up the challenge of "reading that which was never written:"[30] reading, too, *is* a god.[31]

Walter Benjamin proposes a solution to the "enigma," as he calls it, of his inability to recognize people, which may also supply a reason for our collective prosopagnosia, our collective loss of *noos*: "I do not want to be recognized, I want myself to be taken for somebody else."[32] Such a desire to hide, to be mistaken is a clear symptom of shame. In the diagnosis of the psychiatrist, "the wish inherent in the feeling of shame" is: "I want to disappear as the person I have shown myself to be," or: "I want to be [seen as] different than I am."[33] Even more basically:

> "I feel ashamed" means "I do not want to be seen." Therefore, persons who feel ashamed hide themselves or at least avert their faces. However, they also close their eyes and refuse to look. This is a kind of magical gesture, arising from the magical belief that anyone who does not look cannot be looked at.[34]

Rather than magical, or more fundamentally than magical, such a gesture is dictated by our mimetic instinct, which makes us all look for a disguise and warns us that our best chance at being overlooked is by not looking at. In either case, it is an archaic reflex that still dictates our reaction to the face.

Yet we can recognize only if we are willing to be recognized. Only by looking at, we will be looked at in return; only by smiling at, we will be smiled at in return. "A smiling mouth *smiles* only in a human face,"[35] and only, I would add, at another human face. In so doing, however, we become ourselves divine. Virgil's imperative at the end of the fourth eclogue: "Begin, baby boy, to know thy mother with a smile," seems to put the burden of recognition solely on the child, but then we learn that he, "on whom his parents have not smiled,"[36] has been denied intercourse with the gods, namely, both the ability to recognize and be recognized as god-like. Only by being recognized as human, we learn how to recognize the gods. Only when we recognize a human countenance, we are recognized as god-like.

In a note attached twenty-five years later to the *Introduction à la méthode de Léonard de Vinci*, Paul Valéry comments upon a fragment Leonardo probably intended to use in his projected treatise on anatomy: "The organization of our body is such a marvelous thing," Leonardo writes, "that the soul, although *something divine*, is deeply grieved at being separated from the body that was its home. *And I can well believe that its tears and sufferings are not unjustified . . .*"[37] Valéry invites us to consider

> the enormous shadow projected here by an idea in process of formation: death interpreted as a *disaster for the soul*! Death of the body as a diminution of the *divine thing*! Death moving the soul to tears and destroying its dearest work, by the ruin of the structure that the soul had designed for its dwelling![38]

Its mourning shows clearly enough that the soul is not indifferent to the body, to use Leibniz's litotes,[39] but its sorrow is ultimately relieved by the certainty that the separation will only be temporary. Rather than considering such an idea as opposed to the "wholly *naturalistic*" philosophy of Leonardo, as Valéry does,[40] I see in it the culmination of a tradition that goes back to Tertullian. To the initiator of the figural reading[41] the body was certainly no *signum mortificationis* (as the Jesuit Naphta, Hans Castorp's other mentor, would have it),[42] but rather a foreshadowing of the eventual figure of the soul, the face was not a *facies hippocratica* but rather a *veronica* of the coming Messiah. It is by a similar train of thought, I believe, that Emily Dickinson was led to define physiognomy the "strategy of immortality" in one of her most enigmatic poems, the expression I have chosen as the title of this introduction.

Independently from any belief in the resurrection of the flesh, I suggest that such a strategy is most relevant to the battle-field of literary studies.[43] "What is interesting to a writer is the possession of an inward certitude that literary criticism will never die," Joseph Conrad wrote;[44] reversing this disarming declaration of dependence, what is interesting to a critic, I believe, is the possession of "an inward certitude" that literature will never die. To strengthen such an inward certitude is the final goal of my work.

What gives to Proust's art a unique degree of universality "in a non-religious world," according to Adorno, is that he "took the phrase of immortality literally," and he did so "by concentrating on the utterly mortal."[45] Thus, in our fully secularized world, we take the phrase of immortality literally only by taking the letter as immortal, even if only for strategical reasons. I use the word "physiognomy" also to name an approach aware of the unavoidable, yet not unredeemable, materiality of the body and of language.

In the wake of such an awareness, there is hence no reason to mourn the soul, either. Body and soul fall and stand together, as it were. And when we laugh, they also laugh together. Philo of Alexandria has a deeply moving exegesis as to why Abraham falls and laughs at the same time, when God lets him know that Sarah will bring him the child they had so long wished for, now that she is ninety. He explains that Abraham

> falls as a pledge that the proved nothingness of mortality keeps him from vaunting: he laughs to show that the thought that God alone is the cause of good and gracious gifts makes strong his piety. Let created being fall with mourning in its face; it is only what nature demands, so feeble in footing is it, so sad of heart in itself. Then let it be raised up by God and laugh, for God alone is its support and its joy.[46]

It is an eloquent example of Philo's allegorical reading. But Philo refers to his method also as a φυσιογνωμονεῖν, a verb I will not try to translate, as the most recent editors of his works misleadingly do, as "to judge of the real nature of things," but rather simply transliterate as "to physiognomize."[47] His usage suggests that in Philo's eyes the opposition between the literal and the allegorical, which Tertullian tried to bridge with his figural reading, was mediated by a mode of interpretation we may call physiognomical. Benjamin hinted at a similar possibility, I believe, when he listed a "physiognomical criticism" among the future tasks of the critic in his notes for a never completed essay on "The Task of the Critic."[48] I venture to supplement his insight with a formula: the task of the physiognomical critic is to transliterate. In so doing, the critic redresses what is, truly, the failure[49] of the translator and, contrary to the obliteration brought about by translation, furthers the survival of the letter, on which the very survival of literature is dependent.

Jerome answered pope Damasus's inquiry about the meaning of the Hebrew word *hosanna*, which had been left untranslated in the Greek and Latin version of the Scriptures, by arguing that "it is better to accommodate the ear to a foreign-sounding idiom, than to bring home a false understanding of the foreign language (*magis condecet [. . .] peregrino aurem accomodare sermoni, quam de aliena lingua fictam referre sententiam*)."[50] Such a principle would well serve the task of the transliterator, if I may also venture to coin a new word,[51] Jerome might be invoked as their patron saint,[52] his answer to Damasus be their manifesto. It is an example of both common sense and extraordinary humility. Aristotle displays similar qualities in the *Rhetoric* when he justifies the recommendation that "we should give our language a 'foreign air'" by grafting onto it *glōttai*, or foreign words, with the surprisingly enlightened observation that "men feel the same in regard to style as in regard to foreigners and fellow-citizens," namely, "men admire what is remote, and that which excites admiration is pleasant."[53] It is in comedy, according to the Aristotelian author of the tractatus Coislinianus, that every character is made to speak in only one language, without *glōttai*: the countryman in the language of the country, the foreigner in a foreign language.[54] But that is why, transliterating Aristotle, they sound *idiotic*.[55] "La translittération," on the other hand, "a je ne sais quoi de plus intelligent."[56] It reminds us of the unity of all languages, for every word was once a foreign word, before acquiring a familiar physiognomy. It also reminds us that "the secularisation of language is only a *façon de parler*, a phrase,"[57] for in every language, and not just in the Israeli Hebrew that inspired Gershom Scholem's remark, the memory of a divine language is still alive.

In a letter to Scholem, in which he advocates the translatability of Hebrew into German, Franz Rosenzweig comes close to formulate a Freudian theory of the foreign word when he implies that foreign are those words that can *never* come back home: "Worte, die *nie* heimkehren können."[58] Said otherwise, words we can never remember. On the other hand, only those words that can come back home, that we can remember, are, in a Freudian sense, uncanny (*unheimlich*), when they resurface unannounced to memory. Paradoxically, then, one has to conclude that foreign words are not *unheimlich* in a Freudian sense, though they are not necessarily "a nothing, idols (*Götzen*),— אלילים ,"[59] as Rosenzweig denounces them. Such an extreme statement fully reflects the horror of polytheism of the author of *The Star of Redemption*. But I do not believe that the author of *Moses and Monotheism* shared Rosenzweig's disdain for foreign words. Foreign words are not just a nothing to Freud, they rather deserve our attention because, like proper names, they are most liable to be forgotten. In his discussion of the *Psychopathology of Everyday Life*, Freud starts his analysis of the disturbances of memory precisely from proper names and foreign words. As opposed to both categories, what he calls "the current vocabulary of our own language, when it is confined to the range of normal usage," namely, common nouns and the other parts of speech, seem "to be protected against being forgotten."[60] Our own language is, in other words, better protected against forgetfulness than a foreign language, with the exception of proper names. Such a conclusion points toward an affinity between the two categories that are thus excluded from our recollection: both proper names and foreign words are words whose origin *latet*, is hidden from us.[61] If we pursue this train of thought to its utmost consequences, we may attempt to formulate a truly Freudian theory of the foreign word: foreign words are those words we fail to remember, which, in Freudian terms, means: we repress. As Ernst Jones puts it, within Freudian theory "a failure to remember is regarded as synonymous with a not wanting to remember."[62] But what is it exactly that we do not want to remember? We want to repress, I suggest, the embarassing memory that all words, including those we now regard as ours, were once foreign. If every word has once been a loan-word, as Hugo Schuchardt argued,[63] then every word is a forgotten foreign word. To such an extent, every word is *unheimlich*, before we learn to recognize in it a familiar physiognomy.

The difference between ours and other languages then turns out to be mainly an effect of oblivion. Our language is the language we remember. Foreign languages are those languages we have to forget in order to remember ours. Yet we cannot forget that, in order for the various human languages to exist, another memory had to be repressed first: the memory of language as such. The mythical image of the language we forgot is, of course, the language before Babel, its historical image the *lingua franca*, the go-between language Schuchardt rediscovered at the beginning of our century,[64] its messianic image the foreign tongue in which, according to Proust, all "beautiful books are written."[65] But its most memorable image, because at once mythical, historical, and messianic, is Kafka's Yiddish, a language that "consists only of foreign words," which, however, "do not rest within it, but rather preserve the haste and vivacity, with which they were taken in."[66] The convention of printing foreign words in a cursive character wisely visualizes the relationship of the foreign word to its new context, as if the word were ready to jump out of it at will. Like divine names, they have "a life of their own."[67] As Kafka warned his audience, we should cherish the memory of such a language, but without the fear its recollection necessarily carries along.[68] We should not forget, however, that our fear in front of Yiddish is our fear in front of language as such: for language as such is only made of foreign words.

Contrary to Rosenzweig's anathema, therefore, foreign words are not idols, but rather ideas. In a truly Platonic sense, ideas are words we constantly strive to remember, but we keep forgetting. Yet ideas always come back home, they are irrepressible, or, said otherwise, untranslatable, they keep returning out of oblivion in their transliterated form. As he denounces what he regards as the idolatry of foreign words, Rosenzweig picks precisely the word "idea" as the example of a word that never became German, nor can become Hebrew, in spite of Hermann Cohen's patronage[69]—a conclusion certainly neither Benjamin[70] nor Scholem could have endorsed.

The titles of my chapters are meant to recall some of these uncanny words, with no intent, however, of frightening the readers away, but rather the hope of allowing them to savor the joy of recognition,[71] or share in the enlightenment of Plato's *anamnēsis*. In each of them I follow the fate of a number of categories that are, as I show, related to physiognomy, even if they have not necessarily originated within the boundaries of physiognomy as a self-stylized discipline, and have either been borrowed through transliteration, or obliterated through translation, in all the modern European languages.

By recalling these categories by their names, I perform what is, to my mind, the highest duty of any critic: to further the survival of the intellectual vocabulary on which the continuity of our cultural memory ultimately depends.

An overall view of the development of a physiognomical tradition in Western culture underlies my argument. Historically, I argue in chapter 1 "Symmmetry and *Concinnitas*," physiognomy is an offspring of the efforts of ancient medicine to build up a nomenclature of the body parts.[72] Well before laying claim to the status of a hermeneutics of the face, physiognomy was a primer, a spelling book of the body. In order to be able to cast spells on the unwholesome parts of the body, or to dispel the demons infesting it, the ancient physician, who gathered in one person the functions of magician, astrologer, and physiognomist,[73] had first to learn how to spell the body; only much later the need was felt to explain the meaning and the function of each component part of the body in relation to its whole. Plato precisely records, in the very same *Charmides* I have started from, the shift in the diagnostic technique of ancient medicine from a local to a holistic point of view, so to speak. In the opening skirmish, which prefaces the core discussion of *sōphrosynē*, or "health of mind,"[74] Socrates contrasts the healing art of the Thracian shaman Zalmoxis[75] to that of the Greek health-practitioners, who "neglected *the whole* (*to holon*), on which they ought to spend their pains, for if this were out of order it was impossible for *the part* (*to meros*) to be in order."[76] Plato suggests that Socrates uses the anecdote as an allegory of his own therapeutic technique, aiming at healing the body through the soul.[77] When he recommends to Charmides the leaf of an unspecified plant as a remedy against his headache, Socrates is careful to add that the *pharmakon* would not be effective without the simultaneous utterance of what he calls "beautiful words" (*tous logous* [. . .] *tous kalous*), words capable of engendering *sōphrosynē* in the soul. Besides self-referentially introducing the ensuing discussion and legitimating philosophy as a therapy of the soul, this anecdote marks thus a change in the Greek view of the body, as well[78]—from a mere sum of its parts to a whole that amounts to more than just that sum (an organism, rather than a collection of organs).[79] I argue further that mastering the names of the body parts, which was a professional requirement, as it were, of the ancient physician, became a mainly philological endeavor in the Renaissance. The renewed interest in the nomenclature of the body parts was powerfully stimulated by the quest for a recovery of the ancient canon of the human body. The two main categories Renaissance theorists of art rescued from ancient literature and deployed in their attempt to redraw the well-proportioned body were Greek *symmetria* (Latin "proportion"), and Latin *concinnitas* (Greek "harmony"). I discuss the attempt made by authors such as Marsilio Ficino and Leon Battista Alberti to replace *symmetria* with *concinnitas*, and supply a reason for their failure: the Latin body, which translates the Greek and prefigures the modern, is a shamefast body. Neither symmmetry nor harmony, but rather *verecundia* is the bond that holds it together.

In chapter 2 "Character and *Caricatura*" I argue that the uncanny image of ourselves caricature confronts us with is that of our *facies hippocratica*. A change in the ontological status of the body at the moment of death was acknowledged in the ancient cosmos, and reflected by a change in its name, corresponding, namely, to that from "body" to "corpse." The impassive, hieratic figure cast on the face of the corpse by the *rigor mortis*, on the other hand, appeared to the ancients as a token of personal identity and a harbinger of immortality precisely by virtue of the complete erasure of expression it brings about. The imprint of an indelible character, be it at birth or at death, remained through the Christian era the model of a likeness that imitation can at best emulate, but never equal. Only when the repetition of the type was dismissed in the Renaissance as a stylistic failure rather than an assurance of truthfulness, the death mask could be regarded as comic rather than tragic. Caricature is then a prolepsis of death, but its goal is unlikeness, as opposed to the idealized mirroring of portraiture. As it anticipates Kafka's realization that "the light on the grotesque recoiling phiz is true, but nothing else (*Das Licht auf dem zurückweichenden Fratzengesicht ist wahr, sonst nichts*),"[80] the early modern genre of the *caricatura* is, first of all, the parody of ancient character.

The oxymoron "false shame" (and the ensuing conceptual opposition of a vitious to a true, virtuous shame) became commonplace as a translation of Plutarch's *dysōpia* in the Renaissance, when his *Moralia* became again available in the West. "Discountenance" is the translation of the Greek term I propose at the end of chapter 3 "*Dysōpia* and Discountenance." Plutarch explains the word as meaning "to become incapable of facing someone," "to be unable to return somebody's gaze." I argue in favor of the alternative derivation, according to which the face loses its composure and the gaze its directness because of the other's disregard. For this reason, I conclude, "discountenance" is, short of a transliteration, the best approximation to *dysōpia*: somebody's else disregard put us out of countenance. In losing our countenance, however, we do not lose our face, as Plutarch implies, but rather end up gaining one: for the discountenanced is shamefaced. No longer symmetrical, or harmonious, like the Greek, nor decorous, or poised, like the Latin, the discountenanced modern body is a shamefaced body.

In chapter 3 I also start to investigate how the name of an element came to denote in the Italian Renaissance the quality that marks the uniqueness of a human face. In chapter 4, "Air and *Aura*," I suggest that such an extension of its proper meaning was only possible because, before becoming a common noun, "air" had once been a proper name: *aura*. I take my clue from Petrarch's work, and his well-known puns on the name of his beloved, Laura, and argue that, when it resounds anew in his poetry, *aura* revives the fading echo of a divine name. I discuss the usage of the Greek term in the Septuaginta and in the exegetical tradition, especially Philo of Alexandria and the Pseudo-Dionysius, who lists *aura* among the names of God, and trace its occurrences in the Western poetical tradition, up to Dante and Goethe. I am thus able to show that the category of *aura* comes to us from a very different linguistic level than Benjamin thought as he rescued it from those "vulgar mystical books," in which he found it debased to a "halo."[81] Benjamin's famous thesis of a decay of the aura can then be tested at the decay of the word "aura" itself.

The Freudian coinage of the term "prosopagnosia"[82] may suggest that a latent condition affecting our culture, as I have shown, has now become so acute that it can be finally diagnosed. In chapter 5, I introduce two meta-physiognomical categories, as it were: *Nemesis* and *Aphanisis*, on the model of Freud's Eros and Thanatos, to approach the face "expressive chiefly of inexpression" Coleridge foresaw in his own features.[83] I remind us that, before decaying to the status of a common noun, *Nemesis* was also a divine name, and follow its history of transliterations and translations throughout the history of Western culture, up to Freud's translation as the "repetition-compulsion." I oppose to it another category, this time originating in the field of psycho-analysis proper, Ernst Jones's *aphanisis*. Not by chance, the best example of Freud's uncanny is the failure to recognize ourselves we experience when suddenly faced by our mirror-image.[84] As long as we are ashamed of our face, I suggest, we will not be able to recognize ourselves.

Going back to Ostwald's paradox: the eye is indeed "ink-like," but we can only learn to read by overcoming our shame of the ink-like letter. We are ashamed of the letter, as we are ashamed of the face. Yet we should be ashamed, if at all, not of its materiality, but only of its mortality: "not that I am ashamed of the Anatomy of my parts," Sir Thomas Browne writes, "yet I have one part of modesty, which I have seldome discovered in another, that is (to speake truly) I am not so much afraid of death, as ashamed thereof."[85] Immortality is the loss of the shame for our mortality.

Chapter 1

Symmetry and *Concinnitas*

self in self steeped and pashed — quite
Disremembering, dísmémbering áll now.

G. M. Hopkins

Galen had the disconcerting habit, according to a treatise *On the Medical Names* we only know through its Arabic translation, to reply to questions concerning the name of a sickness by throwing back at random proper names, such as: "The name of this fever is Zenon, or Apollonios," or any other name that might come to his mind. Galen justifies the strange practice as a mockery of rival physicians, who were far too interested in the name, rather than in the cause or the treatment of an ailment,

> as if the method, through which one is freed of the fever, would depend on the knowledge of its name and not on the knowledge of the sickness itself and the determination of the things that are necessary in order to treat it.[1]

Even if Galen's polemic is prima facie directed against his contemporaries, his unorthodox method is a parody of the practice of ancient medicine, and a blasphemous parody at that. His calling names mocks the conjuring up by the ancient physician of the divinities that preside over the limbs of the body and are responsible for the sickness affecting the limb they rule. The aching limb is indeed an irated god.[2] Calling the sickness by name was the first task of the physician, the knowledge of the name regarded as vitally important in gaining control over the aching limb at a time when the belief in the magic of language was still unassailed. We may find such a practice less foreign if we consider that even now, as Virginia Woolf points out, when confronting the failure of language to express what he perceives as a uniquely individual pain, the sufferer may be

> forced to coin words himself, and, taking his pain in one hand, and a lump of pure sound in the other (as perhaps the people of Babel did in the beginning), so to crush them together that a brand new word in the end drops out.[3]

Pain, Woolf reminds us, is a stimulus to verbalization rather than to inarticulation.[4] One finds among the preferred techniques of conjuration in ancient Egypt the invocation of each individual part of the body: "there is no part of the body without god," one charm recites, and then proceeds to name and identify each of them with the divinities presiding over them: "each limb is god."[5] But then each disease should be regarded as sacred, the author of the treatise on the sacred disease *par excellence*, epilepsy, does not fail to remark with noticeable sarcasm.[6] In spite of such enlightened criticisms, the practice continued to thrive in Egyptian medicine well into the Christian era. Origen wrote in the third century that the Egyptians divided the human body into thirty-six parts and that each part was under the care of a god. And "by invoking these," namely, the corresponding gods, "they heal the sufferings of the various parts (*ta merē*)."[7]

As the number proves, the partition was supported by the parallel belief in the common ancestor of medicine and astrology, iatromathematics, the astrological healing technique. The zodiacal and planetary *melothesia*,[8] namely, the partition of the human body according to the dominant influences exerted on each individual limb by the zodiacal signs and by the planets, probably evolved as an extension to the human body of the partition of the vault of the sky in Egyptian religion. The sky-goddess Nut naturally encloses the heavenly bodies in her all-encompassing body, whereas the sun-god, when he enters in their dominion on his path, takes on their shape.[9] The advantage of this healing technique over medicine, from the believer's point of view, is evident, when we consider that it does not demand the breaking of the long-lasting taboo of the corpse. It dispenses with the need of autopsy by its system of astral correspondences. The fact that "perhaps the most popular anatomical image during the Middle Ages was the 'zodiac man'"[10] [fig. 2] testifies to the resilience of this magical melothesia well before the beginnings of human anatomy in the Renaissance.

The existence in the West of similar, indigenous beliefs about the body is proven by a famous episode narrated by Livy.[11] As the historian is careful to point out, the apologue by which Menenius Agrippa succeeded in persuading the plebeians to renounce their secession and return to Rome was told in "the quaint and uncouth style of that age" (*prisco illo dicendi et horrido*

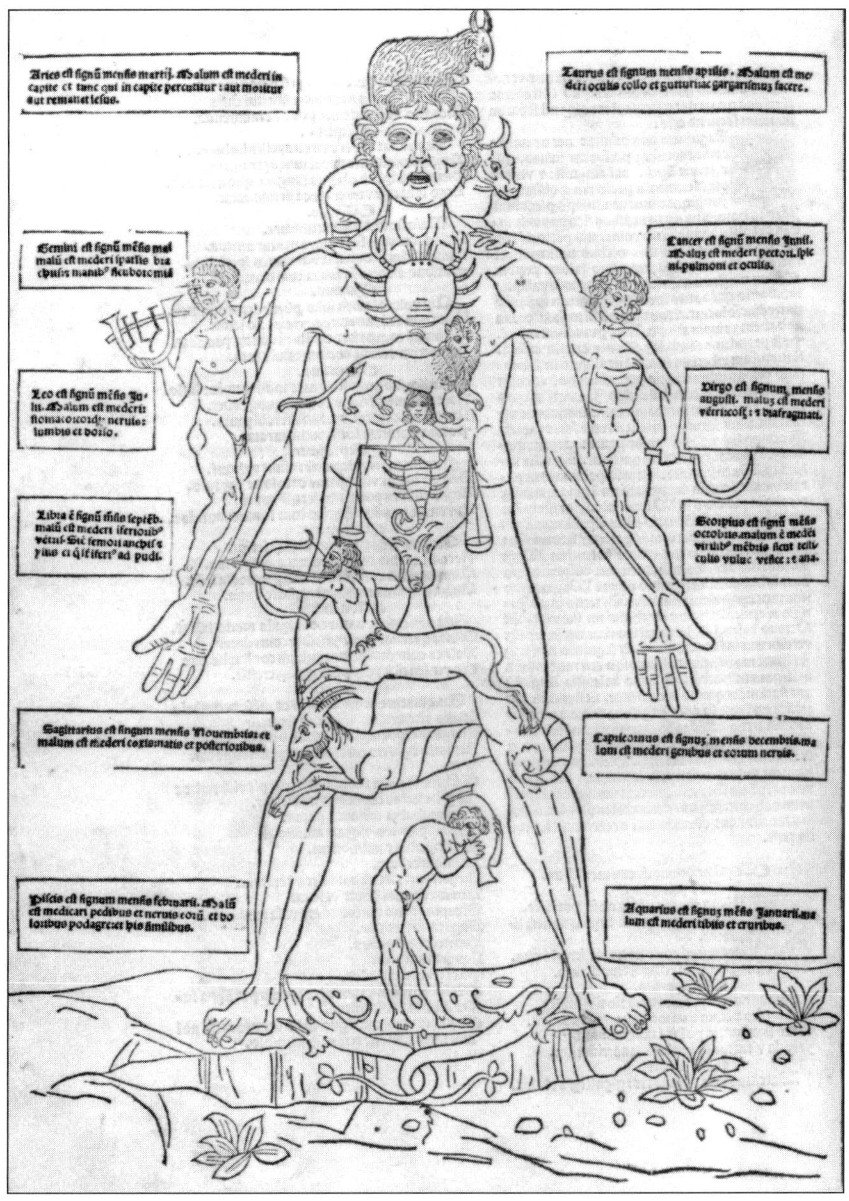

Fig. 2. "Zodiac Man," from Joannes de Ketham, *Fasciculus medicinae,* Venice 1500. Special Collections Research Center, University of Chicago Library.

modo); nonetheless, the Senate ambassador argued eloquently enough the case for the restoration of the social contract. Agrippa manages to restore concord (*concordia*) amongst the citizens by fully resorting to the "fair means or foul" (*per aequa per iniqua*) his mandate entitles him to—he dispatches himself by telling a story:

> In the days when man's members did not all consent amongst themselves, as is now the case, but had each its own opinion and a voice of its own (*tempore quo in homine non, ut nunc, omnia in unum consentiant, sed singulis membris suum cuique consilium suus sermo fuerit*) . . .[12]

Livy can hardly conceal his disbelief at Agrippa's success with so poor a rhetorical device as a straightforward comparison: "Drawing a parallel from this to show how like was the internal dissension of the bodily members to the anger of the plebs against the Fathers, he prevailed upon the minds of his hearers." The ancient physician's mind-set was certainly closer to that of Agrippa's audience than to Livy's enlightened historical sensibility. His re-telling of the anecdote further removes the reader from the possibility of remembering the dismembered body Agrippa so successfully evoked for his audience. In order to understand this view we would have to "disremember"[13] all now, and above all the Pauline rhetoric[14] that no longer allows us to see the body as an unruly collection of parts, each of its own mind, so to speak, rather than a consenting assembly of unanimous members. Pascal codifies with sublime irony the view that still haunts our perception of the body and its component parts, when he writes that "to be a member is to have no life, being, or movement except through the spirit of the body (*être membre est n'avoir de vie, d'être et de mouvement que par l'esprit du corps*)."[15]

The change in the Greek view of the body Plato records in the *Charmides* is a case of the new, general "interest in the relation of the whole to its parts," which "increased especially from the fourth century B.C. onwards" in Greece,[16]—an interest the developments in pre-Socratic medicine and natural philosophy decidedly contributed to awaken. In the *Phaedrus* (269C) Socrates attributes the merit of such an advancement, rather than to an exotic figure such as the shaman Zalmoxis, to Hippocrates himself, who argued that it is impossible to understand "the nature of the body" without considering "the nature of the whole."[17] Hippocrates is here mentioned along with Anaxagoras, the teacher from whom Socrates had expected so much at the time of his youthful infatuation for the "history of nature."[18] On the other hand, a famous piece of intellectual autobiography in the *Phaedon* (96A–98D) assesses the limits of Anaxagoras's as well as of the entire Ionian *physiologia* in rather ironical terms. Socrates describes his disappointment at the discovery that Anaxagoras

did not assign any real causes for the ordering of things, but mentioned as causes air and ether and water and many other absurdities.
And it seemed to me it was very much as if one should say that
Socrates does with intelligence whatever he does, and then, in trying
to give the causes of the particular thing I do, should say first that
I am now sitting here because my body is composed of bones and
sinews, and the bones are hard and have joints which divide them
and the sinews can be contracted and relaxed and, with the flesh and
the skin which contains them all, are laid about the bones; and so,
as the bones are hung loose in their ligaments, the sinews, by relaxing and contracting, make me able to bend my limbs (*ta melē*) now,
and that is the cause of my sitting here with my legs bent.[19]

Whether or not it fairly reflects Anaxagoras's views, this *reductio ad absurdum* confirms that, in Socrates's eyes, enumeration could no longer lend a
proper figure to the human body, and could only amount to a partial account
of the nature of a whole. More importantly, maybe, as it could not explain the
intelligent behavior of Socrates, an even exhaustive nomenclature of the body
parts *a capite ad calcem*, from the head down to the heel,[20] could not provide
the remedy to the insurgence of a disease, either.[21]

In Egyptian medicine the diagnosis had been hardly more than "a verdict
of regional localization of the disease process itself."[22] Such a localization,
however, entails the ability to name the *loci* of the body. In order to name the
disease, it is necessary to identify the part of the body that is hurting. The
Hippocratic physicians still named most conditions after the part of the body
affected, and many of these names, such as *hepatitis*, *arthritis*, *nephritis*, have
survived in current medical terminology.[23] In his study on the genesis of the
Names of the Gods, Hermann Usener chose the many different expressions of
pain that have survived even in the "refined and spiritualized" German language to support his conclusion that a general concept is always a belated
creation preceded by innumerable particular denominations, as personal deities (*persönliche Götter)* are preceded by particular gods (*Sondergötter).*[24]
Following this train of thought, one may conclude that pain always originates
as a local pain before leaving place to an overall concept of "pain" or "disease," and that the expression of pain was always an individual expression of
pain, an individual answer to the individual pain before becoming a diagnostic tool, a universal plea for compassion, or "the general expression of pain"
Reynolds saw on the countenance of Laocoon and his two sons.[25]

The wonder of Plato's Socrates "at that nail of pain and pleasure which
fastens the body to the mind"[26] is then an altogether different reflex from the
awe a piercing pain might have inspired. Firstly, it implies an equivalence of
pain and pleasure that should not be taken for granted; secondly, it assumes
a dislocation of pain that abstracts it from the affected limb and dilutes it over

the entire body. Such is the shift advocated by the author of the Hippocratic treatise *De locis in homine*, who starts by evoking the Heraclitean paradox of the circumference, whose origin can no longer be pointed out, once the tracing thereof has been completed: in the same way, in the body "there is no beginning, but everything is both beginning and end."[27] Hence, stretching the analogy a step further, there is also no beginning (*archē*) to a disease, but every part of the body is both its beginning and its end. The search for a specific pathogenic spot is thus no longer the primary task of the healer: since the parts all communicate with each other, the disease is necessarily transmitted to the entire body, it cannot remain isolated or be isolated for curative purposes.[28] For this reason the therapy, even in an essay drafting an atlas of the body, so to speak, centers around the temporal notion of *kairos* rather than any spatial category. The spreading of the sickness to the entire body leaves few chances to the physician (*hē de iētrikē oligokairos esti*, XLIV.1): what is vitally important is to seize the right moment (*kairos*) for the administration of the remedies. If the right moment is not seized, then the circle is going to close, and the identification of its origin made an impossible and ultimately idle endeavor.[29]

Disease thus becomes a separate entity from the aching limb. Diseases were once "thought to be entirely unlike one another, owing to the difference in their seat (*topoi*)," but the better-knowing author of the treatise *On Breaths* can now pun that, while the *topos* changes, the *tropos* (which one might render here as the course) of a disease is always the same.[30] As a consequence, the body as a whole must now be acknowledged as the site of pain. The wonder is a reaction to the pain, is an attempt at getting rid of the nail. But the body itself must have seemed the nail when the paronomasia *sōma-sēma* could impose itself, and the analogy of the soul with the corpse, sunken in the body as in a grave, could appear enlightening. Loomings of this view are to be seen already in the linguistic usage of the Homeric poems. It has been repeatedly observed, first by the Alexandrian scholar Aristarch, that Homer consistently used the word we use to interpret with "body," *sōma*, in reference to a corpse.[31] The *nomen* of the body is an *omen* of its decline. Plato's first etymology in the *Cratylus* (400C) clearly reflects this knowledge, although the alternative interpretation, of *sōma*/body as *sēma*/sign, already betrays a new interest in the semiotics of the body *per se*, and not just in view of diagnostic purposes.[32] This interest will shortly thereafter result in the development of a physiognomy no longer strictly divinatory, as the one practiced by the Babylonians and probably by them first introduced to the Greeks,[33] nor chiefly prognostical, as we see it applied in the Hippocratic corpus. But to initiate this new practice "a new hierarchy of the passions"[34] was needed,— the hierarchy of the passions that emerged in fourth-century Greece and Aristotle codified in a definitive form in his rhetoric and ethics. As a result, the body could now be valued as a signifier of the passions of the soul in general: "The practice of physiognomy is possible, if one grants that the body and the soul change together, so far as the natural affections go."[35]

The onset of grammatical analysis in ancient Greece shows the influence of the model of medicine and the assumption of the human body as the term of comparison by which to name its elements.[36] The anatomized[37] body provided a most effective model for the analysis of speech into its elements. Grammatical as well as prosodical categories, such as *pous, daktylos, arthron, colon, syndesmos,* all derive from the nomenclature of the body parts;[38] as, more obviously, the actual unities of measurement.[39] The body is taken as the standard by which to measure the universe. Protagoras's noted sceptical saying: "Man is the measure (*metron*) of all things,"[40] if taken literally,—and at least in such a way it was interpreted in the Renaissance—does not read as a relativizing device, but rather leads to the establishment of a standard of truth. The body is the most convenient ordering principle. Alberti, for instance, writes that "all things are learned by comparison (*comparationibus haec omnia discuntur*)," and

> comparison is made with things most immediately known. As man is the best known of all things to man, perhaps Protagoras, in saying that man is the scale and the measure (*modus et mensura*) of all things, meant that accidents in all things are duly compared to and known by the accidents in man.[41]

Since we have standardized unities of measurement, we no longer think of the parts of the body as measurement instruments. "Foot" and "inch" have become for us "dead" metaphors.[42] We associate to the noun an abstract length and not an actual limb. On the other hand, the ancient body is a dimension. It has a waist and a stature.[43] Grammatical categories have fallen prey to a similar forgetfulness. But the very possibility of using the names of body parts in such a special figurative sense ultimately rests on the overall analogy of speech with a living being. The first comparison of the kind occurs in the *Phaidros* (264C), where, in criticizing Lysias's speech, Socrates offers his own philosophy of composition:

> every discourse must be organised (*synestanai*), like a living being, with a body of its own, as it were, so as not to be headless or footless, but to have a middle and members, composed in fitting relation to each other and to the whole (*prepont'allēlois kai tōi holōi gegrammena*).[44]

The term of comparison is here already the new, well-ordered body; more in general, the analogy rests on a new view of the nature of a compound. A compound cannot be accounted for by a simple enumeration: Hesiod's description of a wagon as "a hundred pieces of wood" is not a description definite enough to explain its nature,[45] nor is the syllabification of a name a sufficient account of its etymology.[46] The same word, *syllabē*, is equivocally used throughout the *Theaetetus*, to refer both to syllables in our sense and to any sort of combination. Using such an ambiguity as a leverage, Plato may put forward his own theory that the syllable or combination in general (*syllabē*) is, truly, an "idea" arising out of the several "harmonized" elements (*mia idea ex ekastōn tōn synarmottontōn stoicheiōn gignomenē*), and that the same is true of words and of all other things.[47]

Plato straightforwardly dismisses the archaic view of speech in the *Sophist*, where explicit mention is made for the first time of "the art of grammar" (*grammatikē technē*), understood here literally as the art of properly combining letters (*grammata*) together.[48] He denies the possibility of having a discourse made up of an asyndetic succession of either nouns (*onomata*) alone or verbs (*rēmata*) alone.[49] Yet Plato's grammatical analysis does not cross this threshold; although he stresses in the strongest terms that "the complete separation (*to dialyein*) of each thing from all is the utterly final obliteration (*aphanisis*) of all discourse," and that "our power of discourse (*logos*) is derived from the interweaving (*symplokē*) of the ideas (*tōn ideōn*) with one another,"[50] he limits himself to conclude that a discourse to be such must not merely name, but combine nouns and verbs as its elements.[51] In the *Sophist* Plato uses the term *desmos* in reference to the vowels, which tie together the letters in a word "as a bond," by making the consonants resound;[52] but he does not apply the term to refer to the connecting elements of a sentence. Anticipating the later formal classifications of the parts of speech by the Alexandrian grammarians, Aristotle is the first to have stressed the importance of the connecting elements, which he names generically *syndesmoi*, "ligaments,"[53] for the articulation of meaningful discourse.

The passage from a divinatory to a hermeneutical physiognomy parallels the development of early Greek linguistics, from Plato's onomaturgy in the *Cratylus* to Aristotle's taxonomy of the parts of diction in the twentieth chapter of the *Poetics*. Undoubtedly, a grammatical pattern continued to be operative throughout antiquity in dictating not only the structure of the epic description of beauty,[54] but also the ordering of physiognomical treatises. The anonymous author of the most ancient Latin text in the genre, writing in the fourth century of our era, declaredly follows the order of grammar textbooks, starting with the first elements and proceeding then to combine them as a way of constructing the different types of individuals:

> Since we have properly exposed and enumerated both the signs of
> the limbs and the meanings of these signs, like the first elements of
> the letters, [. . .] let us now conceive and constitute certain types out
> of several of them, as syllables are made out of letters.

And he goes on to construct the type of "the strong man."[55] The composition of a type out of individual features that are previously interpreted in isolation is a standard procedure in most later physiognomical treatises. Such a constructive practice will continue up to the Renaissance and beyond, and produce works such as Giovanni Padovani's *De Singularum humani corporis partium significationibus*[56] or Domenico de' Rubeis's *Tabulae physiognomicae*,[57] which reduce the body to a skeleton-like table of contents, first coordinating to each limb its meaning, and then reassembling them to build up the desired type. Yet physiognomy consistently remains inadvertent of the connecting links, which are throughout its history left out of its scope of interpretation. The number of analyzed limbs remains discrete: alike in this to the sixteenth-century French *blasonneur*, who laments that all the limbs of his lady's body have been already sung, the physiognomist, too, cannot step out of the vicious circle spanned by the same, ever recurring features.[58] There is no physiognomy of the *traits d'union*. The limbs of the physiognomical body are all, as it were, out of joint.

As it emerges from the Middle Ages, physiognomy encompasses, on the one hand, the doctrine of the right construction of the body, which will be later known, following the revival of Vitruvius, as theory of the proportion, or symmetry, of the human body; on the other, the doctrine of the proper mixing of the humors, which determines our temperaments, or theory of the complexions. It aspires to be both a theory of health and a theory of beauty, as Chrysippus had defined health the right proportion of the elements, and beauty the right proportion of the members of the body.[59] The human body is, to the physiognomist, both the meter by which God, the *Primus Mensurator*, as Grossateste calls Him,[60] measures the universe, and the bond (*vinculum*) by which He keeps it together, being "the worthiest of all mixed bodies," as Peter of Abano exaltes it.[61] Later, man as a whole, and not just his body, will be hailed as the "bond or copula of the world" (*nodum et vinculum mundi*),[62] once the Platonic knowledge will be recovered, that "the greatest of symmetries" is "that which exists between the soul itself and the body itself."[63]

Albrecht Dürer's accomplishments as a theorist of art were very early and widely acknowledged, as the rapidly growing European fortune of his writings witnesses.[64] Paolo Gallucci concedes, in the dedicatory letter introducing his Italian translation of the *Vier Bücher von menschlicher Proportion*,—to which he supplemented an influential fifth book on the expression of emotions[65]—that

> Albrecht Dürer [. . .] by far surpassed all those who came before him (even those who are highly celebrated by histories and verses), and left to posterity in his writings and drawings the idea of the true Painting, and of Sculpture, as one can clearly see from his papers, as well as from this book of the symmetry of the human bodies.[66]

But the European resonance of his works occurred almost in spite of Dürer's adoption of his native German. The choice of a vernacular language did not necessarily entail a gain in audience, nor in perspicuity, at such an early stage in the development of the technical vocabulary of art criticism. Gallucci's acknowledgement is startingly qualified on the opposite folio by the sonnet dedicated to him by Girolamo Dandolo:

> *Di Alberto Duro ha in queste carte vita*
> *Il gran dissegno, e del dipinger l'arte,*
> *Mercé di tue virtù, che in ogni parte*
> *Dan spirto à l'opra sua quasi smarrita.*

> Albrecht Dürer's grand design, and the art
> of painting live in these papers,
> thanks to your virtues, which raise
> the spirits of his almost forlorn work.[67]

Probably it is not just for metrical reasons that Dandolo prefers the shortened form "Duro" to the transliteration "Durero" adopted by Gallucci.[68] Dürer's work, the *Vier Bücher von menschlicher Proportion*, first published in German in 1528, would have remained very "hard" (Ital.=*duro*) indeed to its readers, had it not been translated into Latin by his friend, the humanist Joachim Camerarius. Published under the title *De Symmetria partium in rectis formis humanorum corporum*, Camerarius's "splendid translation" was even at the time, as Erwin Panofsky has observed, "indispensable for the understanding of Dürer's archaic German,"[69] and thereby essentially contributed to the European reception of his work. But the difficulty of Dürer's language is foremost due to his need to invent ex novo a terminology for naming the limbs of the body with painstaking precision. Gallucci remarks with admiration in his preface that Dürer

> did not leave any small exterior particle of our bodies (for the painter
> and the sculptor consider nothing else in man than that which is
> seen) unmeasured and unexplained by his divine mind, with such a
> subtlety that astonishes all lovers and experts of art.[70]

Camerarius was very much aware of the difficulty of his task in taking on such an ambitious *tour de force*, and discussed it in detail in the introduction to his translation. There he writes:

> the author sought out with an almost uncanny diligence names for
> the parts of the human body wherewith the measurements could be
> made more precise, and to some of them even imposed new names.
> I trust that the students of my version will understand this additional

difficulty, not to mention all the other problems I met with, which cannot seem easy to solve, given that there is nothing imitable in this genre from antiquity. However, we divised with no mediocre effort nor little time names wherewith we could render Dürer's own, and we leave up to the readers to decide if they are appropriate.[71]

He lets this *caveat* be followed by an overview of his choices, a "tabula rationum," listing side by side Dürer's German word and his own Latin interpretation, sometimes along with the corresponding Greek term. For instance:

Sinciput graeci vocant βρέγμα. *intelligenda est capitis summitas pro qua veteres verticem posuere.* Die Scheytel.

The Greeks call the crown of the head βρέγμα.

Whereby they mean the apex of the head, what the ancient called vertex. *Die Scheytel.*

Jugula et juguli. Halsgrüblein. *Intelligenda autem est* ἡ σφαγή, *id est lacuna in mediis jugulis.*

The throat. *Halsgrüblein.* It means ἡ σφαγή, namely, the throat-pit,[72]

and so forth. Camerarius's interest in the topic, which has crystallized itself in his *Commentarii utriusque Linguae*,[73] a bilingual dictionary of the body, a signal achievement of humanistic erudition, was most likely stimulated by his work at this translation. But the quest for the proper names of the parts of the human body is a humanistic endeavour *par excellence.*[74] Nomenclature is "an important matter at the period when the naming of anatomical parts was influenced variously by Greek, Latin, Arabic, and Hebrew terms;"[75] it is also a main concern of early modern physiognomy. During the course of the sixteenth century, physiognomy becomes a more and more philological discipline, to the point that it might be legitimately named an "Anatomia Philologica," borrowing the title of a 1632 treatise by Gregorius Queccius.[76] On the one hand, it aims at a restitution of the proper names for the parts of the body in the classical languages, on the other, at their correct translation in the various vernacular idioms.[77]

As a consequence of this process the humanists gathered new evidence of the richness of the Greek language,—Camerarius writes with admiration in his *Commentarii* that the Greeks, "a nation rich of words," left no part of the body without a name[78]—; but they were also reminded of the comparative poverty of Latin,—the *egestas* Lucretius had first lamented—especially in the borderline domain of ethopoeia, or character description. Willibald Pirckheimer, the Nürnberg humanist, writes in the dedication of his translation of Theophrastus's *Characters* (1527) to his friend Dürer: "in translating some expressions, I could not even satisfy myself, that which occurred not for a fault of mine, but rather for the poverty of the Latin tongue."[79]

The inadequacy of the Latin language had to be particularly felt in the case of those Greek words the Latins themselves had declared untranslatable and directly transliterated in their own tongue. Such is the case of "symmetria," the term Camerarius chooses to translate Dürer's "Proportion." The word has since been borrowed by most modern European languages, but is now used in a largely different way, to refer almost unambiguously to bilateral symmetry.[80] Camerarius writes with a polemical overtone in the same introduction I have already cited: "Even if they deny to have a Latin name for it, we will nevertheless translate it as *commensus* or *commensuratio*." (*Latinum nomen etsi habere negant nos tamen interpretemur commensum commensurationemve.*) Naturally Camerarius very well knew—in spite of Pliny's denial: "non habet latinum nomen symmetria" (*Nat. Hist.* 34, 65)—that his calques were no neologisms. Vitruvius himself had used "commensus" for "symmetria" (III.i.2); and "commensuratio" is also attested in post-classical Latin for the same purpose.[81] Yet by using the Graecism *symmetria* instead of transliterating Dürer's *Proportion* backward into Latin, Camerarius was able at once to pay homage both to the Greek tongue and to the authority of Cicero and Vitruvius. Cicero had tentatively proposed the neologism "proportio" in his version of the *Timaeus*; but as a translation of *analogia*, and not of *symmetria*.[82] And Vitruvius had articulated the relationship between the two terms in the following passage, which opens the third book of his treatise: symmetry

> arises from proportion (which in Greek is called ἀναλογία). Proportion consists in taking a fixed module, in each case, both for the parts of a building and for the whole, by which the method of symmetry is put into practice. For without symmetry and proportion no temple can have a regular plan; that is, it must have an exact proportion worked out after the fashion of the members of a finely-shaped human body.[83]

The passage created many problems to those early translators, who tried to render Vitruvius's technical idiom into their vernaculars. The first Italian translation by Cesare Cesariano certainly does not make it any clearer:

> questa [la symmetria] si aparturisse da la proportione: quale graecamente analogia si dice. La Proportione si e de la rata parte de li membri in ogni opera & del tuto la commodulatione. da la quale si effice la ratione de le symmetrie. Imperoche non po alcuna aede senza symmetria & anche proportione habere la ratione de la

> compositione: se non como al imagine de uno homo bene figurato
> de li membri hauera auto exacta la ratione.[84]

It is evident that the translator thought it best to circumvent the difficulties the text presents by simply molding his own language on that of the original, with almost comical results. It has been written apropos of this translation that

> Cesariano, even if he wanted to, could not write in vernacular. Some-
> times, out of despair, he resorted to Latin, but on the whole his effort
> as translator and commentator was useless. For the language he
> pretended to write in could not be that which he spoke, the Lombard
> dialect; nor a language a layman could oppose, as free and loose
> from any rule as it still was, to the pressure of a difficult Latin text.

The critic concludes his quite harsh review with the remark that "only today our historical curiosity and philological expertise may patiently unseal the text."[85] Thus one fails, however, to take into account the problematic status of Vitruvius's text itself, whose Latin is not at all better off in dealing with complex Greek concepts. Alberti observes in the opening paragraphs of the sixth book of his *De re aedificatoria*, certainly with a hindsight *pro domo sua*, that Vitruvius

> wrote in such a Manner, that to the *Latins* he seemed to write *Greek*,
> and to the *Greeks*, *Latin*: But indeed it is plain from the Book itself,
> that he wrote neither *Greek* nor *Latin*, and he might almost as well
> have never wrote [*sic*] at all, at least with regard to us, since we
> cannot understand him.[86]

The complaint is echoed by Francesco di Giorgio, who prefaces his incomplete translation of Vitruvius's treatise by lamenting that "by virtue of Greek and Latin scholarship it has never been possible to master such a task (*per forza di grammatica greca e latina non è stato mai possibile venirne al fine*)."[87]

However, the survival of *symmetria* in transliteration suggests that the difficulty here lies well beyond the shortcomings of the individual translator. No translation has been able to replace the word, which has passed from one language to another, while keeping all its ambiguity in the process. Yet many attempts were made to decode it, especially when the recovery of the forlorn *symmetria prisca*[88] seemed to lie at hand. In the "Proemio" to his commentary on Dante's *Comedy*, Cristoforo Landino offers a brief overview of the development of the figurative arts in Florence, which anticipates Vasari's standard treatment. There he attributes to Cimabue the merit of reviving painting, which had for centuries produced "dead" figures, "unsuited to display any affection of the soul (*punto atteggiate e sanza affetto alcuno d'animo*)," by rediscovering the "true proportion, which the Greeks call symmetry (*vera proporzione, la quale e' Greci chiamano simetria*)."[89] The rediscovery of the "true proportion" of the ancients also entails the recovery of the "true" meaning of the name "symmetry." But its mere mention could not satisfy all those interpreters who, at a very early stage in the renaissance of Greek studies in the Western world, shared Bruni's conviction that "there is nothing said in Greek, which cannot be said in Latin (*nihil graece dictum est, quod latine dici non possit*)."[90] The Pavia humanist Giorgio Valla, one of the most prolific translators of the Renaissance,[91] proposes his own interpretation in the widely read encyclopedia *De expetendis, et fugiendis rebus*: "Symmetria [. . .] latine commensurabilitas dici potest."[92] But already Pomponius Gauricus, although he largely relies on Valla's erudition for the chapters on "symmetria" and "physiognomonia" in his treatise *De statua*,[93] shows his dissatisfaction with this choice by using "commensuratio" or even "mensura" in its stead;[94] while he picks "commensus" for "analogia" and rebuffs Cicero's choice of "proportio" for the same term.[95] Cesariano himself ventures to paraphrase *symmetria* as "numeratione commensurabile."[96] In a different context, that of Scaligero's *Poetics*, the choice falls on "convenientia."[97] Examples of this sort could be multiplied. A critical assessment of modern Vitruvian versions, to the effect that "each author translates the different passages differently,"[98] clearly applies to earlier attempts, and to the different words of this elusive text, as well. Yet can we blame the failure of the individual translators, or their disagreement, for what has been the historic outcome of this diatribe, namely, the simple transposition as a loan of this category into all the modern European languages? In other and more general terms: should we consider the loan of a word the acknowledgment of a subjective failure to understand, or the result of an objective untranslatability?

Schuchardt's paradoxical principle: "jedes Wort ist irgend einmal ein Lehnwort gewesen"[99] offers maybe a way out of this only apparent alternative. If all words have once been loan-words, all translation has once been transliteration.[100] Συμμετρία had first to become *symmetria* in order to become "symmetry." In this—all but automatic, yet irreversible—transition, meaning was lost. The word "symmetria" met thus the same destiny most words of our intellectual vocabulary fell prey to: thought abandoned them to speech;[101] yet speech kept them alive. Words survive thought as living elegies to what they once signified.[102]

Translation is "the death of understanding"[103] because it is the death of the letter. Transliteration is its transfiguration. Translation killeth, but transliteration giveth life. Transliteration is the movement that counteracts the obliteration of the letter brought about by translation. Transliteration, and not translation,—not even, as Benjamin would like, the interlinear version of the Scriptures, which to him represents "the prototype or the ideal of all translation" (*das Urbild oder Ideal aller Übersetzung*)[104]—harbingers the survival of the original, its immortality, or just the eventuality of a revival; even if at the price of its immediate understandability. The dismaying outcome of Averroes's search[105] for the meaning of the words "tragedy" and "comedy" remained thus harmless: his misunderstanding did not curtail their survival.

Such a conclusion might sound less paradoxical if we consider that, instead of translating, we are always transliterating; or at least we do so in all Romance languages. The very word Italian and all the other Romance languages use for "translation" literally means "transliteration." The Latin verb *traducere* Aulus Gellius uses to refer to the transport of Greek terms into Latin was taken by the Italian humanist Leonardo Bruni to mean "translation" and as such passed into the Romance *koine*. Gellius writes "vocabulum graecum vetus traductum in linguam romanam." Leonardo Bruni misunderstood the term as meaning "translated," whereas, as the context makes clear, it meant, literally, transliterated.[106]

The definition of symmetry in the opening chapter of the third book of the *De architectura* is followed by the Vitruvian theory of the proportions of the "homo bene figuratus." Discussed and illustrated with relentless interest throughout the Renaissance [fig. 3], Vitruvius's canon has been the point of departure of all later attempts to codify anthropometry, [107] as well as of all those aesthetic theories of the Renaissance that interpret beauty as "Vergleichlichkeit." This is Dürer's own term of choice for "symmetry."[108] However, in the Vitruvian lexicon he dedicated to his patron Markus Welser, Bernardino Baldi writes that Germans translate "symmetria" as "rechtmessigung" and "gleichförmung," namely, "rectum seu continuum commensum, et similem deformationem."[109] Baldi also writes that "nos Itali *proportionem et correspondentiam* dicimus."[110] As we have seen, the choices available to the Italian interpreter were even more numerous; and since Alberti had rescued the term from Cicero, yet another translation was at hand: *concinnitas*.[111]

In the treatise *On Painting* the word occurs in the Latin text only, without a counterpart in the Italian version. *Concinnitas* is here supposed to result from a fitting composition of surfaces, whereas *symmetria* (which is also missing in the Italian version) is the result of a fitting composition of members.[112] In the later *De re aedificatoria concinnitas* resurfaces as a central category of Alberti's aesthetics. He uses it to define "pulchritudo":

> I shall define beauty to be a Harmony (*concinnitas*) of all the Parts, in whatsoever Subject it appears, fitted together with such Proportion and Connection, that nothing could be added, diminished or altered, but for the Worse.[113]

The eighteenth-century translator James Leoni chooses here "harmony"; then in the tenth book, where a different definition is given, which no

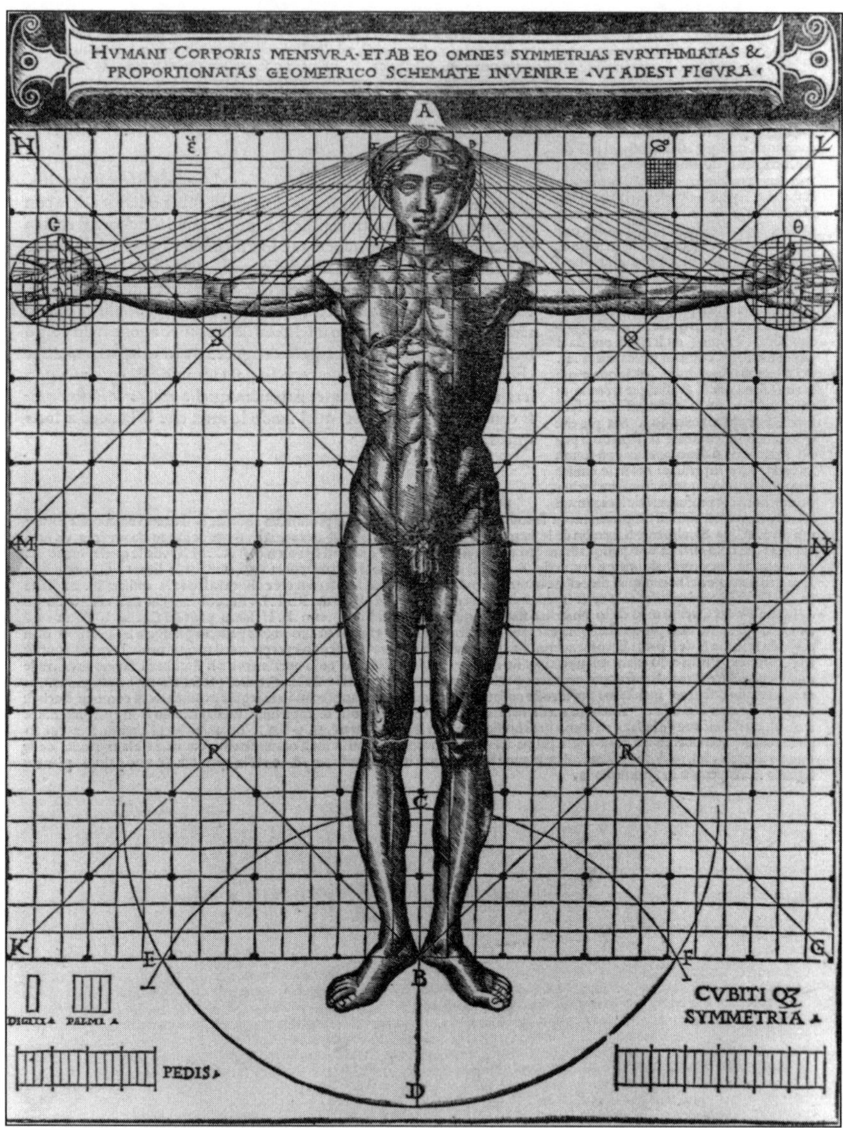

Fig. 3. "Homo bene figuratus," from *Di Lucio Vitruuio Pollione de Architectura Libri Dece traducti de latino in Vulgare,* Como 1521. Special Collections Research Center, University of Chicago Library.

longer straightforwardly identifies beauty with *concinnitas*, Leoni interprets it as "congruity":

> we may conclude Beauty to be such a Consent and Agreement of the Parts (*quendam consensum et conspirationem partium*) of a Whole in which it is found, as to Number, Finishing and Collocation, as Congruity, that is to say, the principal Law of Nature (*concinnitas, hoc est absoluta primariaque ratio naturae*) requires.[114]

As Leoni's oscillation betrays, the word has not been assimilated by the English language,[115] nor by Italian, even in spite of Alberti's and Ficino's patronage. Ficino uses it in his commentary on Plato's *Symposium*, but only in the Latin version, where, like Alberti, he invokes the category to define beauty:

> Beauty is, in fact, a certain charm (*Pulchritudo vero gratia quaedam est*) which is found chiefly and predominantly in the harmony of several elements (*in concinnitate plurium.*) This charm is threefold: there is a certain charm in the soul, in the harmony of several virtues (*ex plurium virtutum concinnitate*); charm is found in material objects, in the harmony of several colors and lines (*ex plurium colorum linearumque concordia*); and likewise charm in sound is the best harmony of several tones (*ex vocium plurium consonantia*).[116]

In the Italian version of the passage Ficino simply transliterates *concordia* and *consonantia*, while consistently rendering *concinnitas* with "conrispondentia"; whereas for the other occurrences of the word in the treatise he uses throughout "consonantia."[117] In recording Ficino's definition in his treatise *On the Beauty of Women*, Firenzuola feels the need to explain Ficino's unfamiliar usage, and does so while transliterating the word into Italian, against the author's own example:

> Ficino, the Platonist, in his work on the *Symposium*, in the second oration, says that beauty is a certain grace that comes from the concise union of several parts; and he uses the term concise because it implies a sweet and charming order, something akin to an elegant collective (*la bellezza è una certa grazia, la quale nasce dalla concinità di più membri: e dice concinità, perciocché quel vocabolo importa un certo ordine, dolce e pieno di garbo, e quasi vuol dire uno attillato aggregamento*).[118]

The most recent translators of Firenzuola's text proceed just by way of assonance when they translate *concinnità* as "concision." As Ficino's usage shows quite clearly, for him *concinnitas* was quasi-synonymous with *consonantia*. *Consonantia*, English "consonance," is directly calqued on Greek *symphonia*, but the adjective *concinnus*, from which *concinnitas*, is also explained as *symphonos* by the ancient glossarists. Priscianus interprets *concinnus* as a compound from the verb *canere*, "to sing."[119] The meaning of the word would hence be closer to *concentus*, from which our word "concert."[120] But another, more likely explanation was advanced by Nonius Marcellus, according to which the verb *concinnare* would derive from *cinnus*, the ancient name of a drink made out of various beverages.[121] Such a derivation relates the word *concinnus* to another field of knowledge, the theory of the complexions, in Greek *krasis*, a word that also referred to a mixture of beverages.[122] *Concinnus* would hence mean "well-mixed," from which the extension to the field of music in the sense of "well-tempered" would have been quite effortless.[123] It is probably to be assumed that this derivation was then forgotten in favor of the more obvious one, assimilating the two verbs *concinnare* and *concinere* in popular etymology.[124]

As in the case of the numerous Latin calques on *symmetria* that were ventured to substitute it, the word *concinnitas*, too, has not been able to supplant its Greek ancestors. In Plato's *Symposium* the physician Eryximachus, who not by chance is charged of putting forth an interpretation of beauty as harmony, equals *harmonia* and *symphonia*: "harmony is consonance, and consonance is a kind of agreement (*hē gar harmonia symphonia esti, symphonia de homologia tis*)."[125] As the translation I quote shows, it is preferable and, indeed, necessary to replace the second Greek word with its Latin calque, for the word "symphony" has now taken a much more limited technical meaning, to refer to a specific musical form, and only rarely can be used instead of its Latin alias. The word "harmony," on the other hand, has kept a wider semantic range, and is not just limited to the musical realm. Few years after Ficino, Pico could write that "the word 'harmony' in its general sense can mean the normal state of order in any composite thing," although "strictly speaking it means only the arranging of several notes which fit together to make a pleasant sound."[126] But the semantic range of "harmony" extended well beyond music already in Greek. Before becoming a musical term, the word referred to any kind of fitting together. The words "art," "rite," "article," "articulation," all derive from the same Indoeuropean root **ar-*, from which "harmony" derives.[127] The use of "harmony" in reference to an invisible attunement is hence a catachresis, and not vice versa.[128] Aetius's authority, according to which Pythagoras, who coined the two terms, used *symmetria* and *harmonia* as synonymous,[129] comes thus as the belated mythical explanation of a fathomed affinity. Symmetry is the mirror-image of harmony, harmony the echo of symmetry. But how did the word "harmony" outlive the "untuning of the sky," the obsolescence of both the Classical and Christian theories of world harmony,[130]—as symmetry outlived the decanonization of the human body?

Kant justifies his defense of the philosophical viability of a term such as *idea*, which has, of course, also survived in transliteration in all the modern European languages, by arguing that "to coin new words is to advance a claim to legislation in language that seldom succeeds."[131] If not the ideal, certainly the most *legitimate* form of translation, one will be entitled to conclude, is transliteration. In one of his postumously published marginalia, Kant transcribes an etymology current in the eighteenth century, according to which words have a value as money does: "verba valent sicut numi," where the Latin name *numi* is made to derive from the Greek *nomos*, "law."[132] *Nomina* are *numi*, names are a currency whose value cannot be altered at will, but is prescribed by law. However, by extension of its proper meaning, the word *nomos* means also "song," and Aristotle, to whom the former etymology goes back, explained the coincidence as due to the circumstance that the ancients, not having any writing, sang their laws "to avoid forgetting them."[133] Before becoming liable to interpretation, the letter of the law had to be taken to heart.

Along with his rewording of Ficino's definition, Firenzuola lists those theories of beauty that were battling the ground with the champion of Platonism at this stage of the Italian Renaissance:

> In his *Tusculanae*, Cicero says that beauty consists of a suitable arrangement of parts with a certain softness of color (*la bellezza è un'atta figura de' membri, con certa soavità di colore*). Others, one of whom was Aristotle, said it is a certain appropriate proportion arising from the manner in which differing parts go together one with the other (*una certa proporzione conveniente, che ridonda da uno accozzamento delle membra diverse le une dall'altre*).[134]

Even if Cicero's definition is more immediately inspired by Chrysippus,[135] both can be considered versions of what Benedetto Varchi calls Aristotelian, or corporeal beauty;[136] as the fact that Ficino, when attacking these antagonist views, combines and criticises them jointly, may also confirm: "there are some who think that beauty consists in a disposition of parts, or, to use their own language, size and proportion together with a certain agreableness of colors."[137]

Yet Aristotelian beauty, even in its Stoic version, is ultimately rooted in the Pythagorean tradition. Galen's treatise on the temperaments, which is our main source on the issue, links Chrysippus's theory of symmetry to its archetype, Polycleitos's canon. According to Galen, the famed sculptor and disciple of Pythagoras codified corporeal beauty once and for all, in a truly, *per antonomasiam*, canonic way: having taught all the proportions of the body, he thought well to support "his treatise with a work; he made a statue according to the tenets of his treatise, and called the statue, like the work, the

'Canon,' which got such a name from having precise commensurability (*symmetria*) of all the parts to one another."[138] This statue is obviously the prototype of the Vitruvian "homo bene figuratus."[139] Ancient artists, according to Pliny, used this statue "to draw their artistic outlines [. . .] as from a sort of law."[140]

In addition to their decisive contribution to the canonization of the body, the Pythagoreans were also credited for upholding the theory of the soul as harmony of the body parts, which Simmias famously defended in the *Phaedon*.[141] Cicero writes in the above-mentioned *Tusculan Dialogues* that the Pythagorean Aristoxenus

> held the soul to be a special tuning-up of the natural body analo-
> gous to that which is called harmony in vocal and instrumental
> music; answering to the nature and conformation of the whole
> body, vibrations of different kinds are produced just as sounds are
> in vocal music.[142]

Later Christian interpreters tried to spiritualize this theory by suggesting that harmony might be the bound uniting a separately existing soul to the body; but this against the evidence provided by all the ancient interpreters.[143] If the Pythagorean theory is hence ultimately materialistic, one might conclude that the Aristotelian, or, corporeal beauty is nothing else but the harmony of the body the Pythagorean soul no longer is. An aesthetic theory replaces a psychological theory. Beauty replaces the soul as the harmony of the body, and lingers in its stead over the features of the face. Beauty is the (material) soul of a body.

Firenzuola goes on to offer his own compromise solution, which tries to reconcile the theory of beauty as visible symmetry with the theory of beauty as audible harmony. He studiously avoids, thus already confirming its ephemerality, *concinnità* in favor of "harmony," although, like Pico, he feels the need to justify its employ of the latter term in reference to visible beauty:

> beauty is nothing else but ordered concord, akin to a harmony that
> arises misteriously from the composition, union, and conjunction of
> several diverse and different parts (*una ordinata concordia, e quasi
> un'armonia occultamente risultante dalla composizione, unione, e
> commissione di più membri diversi*) that are, according to their own
> needs and qualities, differently well proportioned and in some way
> beautiful, and which, before they unite themselves into a whole, are
> different and discordant among themselves. I have said concord and
> harmony as if by way of a simile [*Dico concordia, e quasi armonia,
> come per similitudine*], for just as in music the concordance of high,
> low, and other voices produces the beauty of vocal harmony, so too

> a stout limb, a thin one, a light one, a dark one, a straight one, a
> curved one, a little one, a big one, arranged and joined together by
> Nature in an inexplicable relationship (*con un incomprensibile
> proporzione*), create that pleasing unity, that propriety, that modera-
> tion we call beauty (*quella grata unione, quel decoro, quella
> temperanza che noi chiamiamo bellezza*).

He still pays an implicit homage to Alberti's identification of *concinnitas*
with a "principal law of nature" when he concludes that beauty "can only
come from a mysterious order in Nature (*uno occulto ordine della natura*),"
but retorts that "in my opinion, the human intellect cannot fathom" such
an order.[144]

The members of the body are arranged according to rules of composition
that are dictated by nature. Physiognomy, as the most credible of its etymologies suggests (from *physis* and *gnōmē*, "rule of nature"), is meant to spell out
the hidden rules of composition of the human body, those rules that determine
the make-up of each individual.[145] The Italian word *leggiadria*, if taken in its
own etymological import (from *legge*, "law"), provides a possible approximation to the Greek. Firenzuola defines it in his treatise following the definition
of beauty I just quoted:

> according to some, and to what the word itself says, elegance
> (*leggiadria*) is nothing more than the observance of an unspoken
> law, given and promulgated by Nature [. . .] for the movement, bear-
> ing, and use both of your entire body and of your specific limbs with
> grace, modesty, gentility, measure, style, so that no movement, no
> gesture, be without moderation, without manner, without measure,
> without intention, but rather, as this unspoken law obliges us, it be
> trimmed, composed, regulated, graceful. Because this law is not
> written down anywhere but in a certain natural judgement which of
> itself neither knows nor can explain the reason, except that Nature
> wants it like this, I have called it unspoken.[146]

The observance of the law of nature, according to Firenzuola, inadvertently
graces with beauty, and lawfulness spontaneously turns to *leggiadria*. The
law gratuitously bestows, as it were, airs and graces on its subjects. The shift
from lawfulness to beauty Firenzuola describes, however, is all but natural. It
is rather the recounting of the epochal shift from the pre-historic, unruly
body, to the historic, harmonious body. This transformation demanded the
allegiance of each individual limb to the covenant bringing them together
under a common law. As Livy implies,—and this he suggests to be the ultimate moral of Agrippa's story, when viewed in the larger frame of his history
ab urbe condita—conviction does not suffice to lay the groundwork of a

commonwealth, since history teaches that "Covenants, without the Sword, are but Words." Society is at most a sonorous, not a harmonious body. By pacts and covenants, Hobbes writes in the introduction to the *Leviathan*, the most spectacular and extensive application of the analogy between the body natural and the body politic, "the parts of this Body Politique were at first made, set together, and united; hence they resemble that *Fiat*, or the *Let us make man*, pronounced by God in the Creation."[147] But another divine artificer reminds his audience of vicarious gods, in Plato's *Timaeus*, that "all that is bound may be dissolved,"[148] were it not for His will that holds the whole together. The archaic body had to be subject to the rule of law in order to develop into a well-tuned body.

Talking about composition, or *synthesis*, which he defines as a *harmonia* of words, the author of the treatise *On the Sublime* compares it to the *system* of the human body: "None of the members has any value by itself apart from the others, yet one with another they all constitute a perfect system (*systēma*)."[149] But this standard of comparison was not yet a canon, at least as long as the theoreticians of composition acknowledged the existence of different types of harmonies, each endowed with a different "character" (*charactēr*), "as in personal appearance (*opsis*), so in literary composition."[150]

In 1555 the physician Jean Lyege published in Paris a poem in hexameters in four books under the title *De humani corporis harmonia*. The poem is preceded by a note to the reader, in which the author tries to excuse himself for his "rather hard verses" (*duriusculos versus*),[151] by claiming that he had just aimed at rendering faithfully the medical terminology rather than at achieving a harmonius versification. Undoubtedly, the names of the members of the body do not seem to fit the "hard" harmony[152] of the verses of this unlikely Lucretius. The poem opens as follows:

> *Corporis humani partes, potioraque membra,*
> *Cumque usu formas horum, numerumque situmque,*
> *Multiplices motus, concinnas denique moles,*
> *Versibus expedio medicae fautoribus artis*;

which, made even more prosaical, sounds:

> The parts of the human body, and its major members,
> their use and their shapes, their number and site,
> their multiple movements, and concinnous masses,
> I compose in verses destined to the physicians.[153]

Yet by attempting to fit the human body within the cast of a verse, Lyege was able to indulge both his vocations, as poet and physician. From the eponym of Western poetry[154] onward, the task of the poet has always been, in a fundamental sense, that of harmonizing the body in pieces, of rebuilding its scattered members in the unity of a verse—and thus of reversing, so to speak, the tendency of the body to loosen itself, to decompose. Poetry recomposes the *disiecti membra poetae*, if only, at first, on an imaginary level and by way of enumeration: in it "a hand, a foot, a face, a leg, a head," already stand "for the whole to be imagined."[155] The body is remembered out of its dismembered limbs, if only, at first, in a purely metonymical way. In a more literal sense, each line of a poem was once a lineament of the body. The Greek *melos*, from which our melody,[156] is a singularization of the Homeric expression *melea*, used only in the plural to refer to the members of the body.[157] But Parmenides's *krasis meleōn* is, already no longer in a Homeric sense, both the well-tempered juncture of the bodily members and a certain melodic structure.[158] The epic body differs from its epigone. The master trope of the epic body is the asyndeton,[159] whereas the epigonal body corresponds rather to the scheme of a polysyndeton. In the above-mentioned *Commentarii* Camerarius interprets *sōma* as "membrorum apta compositio et concinnatio,"[160] a description that is certainly anachronistic when applied to the body of the Homeric heroes.

On the other hand, if not beauty, enumeration could certainly bestow sublimity on the archaic body. Longinus praises the asyndeton as an element of the sublime and dismisses the usage of *syndesmoi*, or conjunctions, which frame the expression of emotions in an unnatural slow motion:

> if the rush and ruggedness of the emotion (*to pathos*) is levelled and smoothed out by the use of connecting particles, it loses its sting and its fire is quickly put out. For just as you deprive runners of their speed if you bind them up, emotion equally resents being hampered by connecting particles (*syndesmoi*) and other appendages.[161]

Discourse is a kind of running, according to the etymology of the Latin name (*dis-cursus*).[162] The archaic body is a discursive body, the articulated body that replaces it is, at most, a digressive one, a slow pacer. As such, it is better capable of controlling its emotions and of taming them into habits, whereas the austere harmony of the archaic body is better suited to portray *pathos*.[163]

Once again, the transition is best perceived at the decisive caesura of the translation of Greek into Latin. In a letter concerning the proper style for a philosopher Seneca recommends to his pupil Lucilius not to imitate the Greeks, who like to indulge in an unrestrained style, whereas the Romans have become accustomed to the use of signs of interpunction even in writing.[164] But Seneca's main objection to a discursive practice is moral, and not stylistic, for the price to pay for speed of speech is a loss of shame: "you could only be successful in practising this style by losing your sense of modesty (*si te pudere desierit*); you would have to rub all your shame from your countenance (*perfrices frontem oportet*)."[165] The Latin body, which translates the Greek and prefigures the modern, is, at first sight, a shamefast body. *Verecundia*, and no longer symmmetry nor harmony, is the bound that now holds it together.

Chapter 2

Character and *Caricatura*

There are hardly any two things
more essentially different than
character and *caricature*.

William Hogarth

In Hippocrates's treatise on prognosis, the body is still a foreboding of its own inevitable decay, and the face the foremost display area of its symptoms. The first recommendation to the physician is to examine the face of the patient, and especially to notice "whether it is like its usual self. Such likeness will be the best sign, and the greatest unlikeness will be the most dangerous sign." Then the description follows of what has come to be known, by antonomasia, as *facies hippocratica*: "Nose sharp, eyes hollow, temples sunken, ears cold and contracted with their lobes turned outwards, the skin about the face hard and tense and parched, the colour of the face as a whole being yellow or black."[1]

Unlikeness to oneself is the most mortal sign, because it prefigures the loss of individual identity brought about by death.[2] Likeness, on the other hand, is a visible proof of continuous identity and a token of immortality: hence the cult value of practices meant to immortalize the body, such as mummification and portraiture, in Egyptian religion and beyond.[3] Hamlet's question, "How long will a man lie i'th'earth ere he rot?" still betrays the same anxiety about the status of the inhumed body. (The Egyptian in us, one[4] would say, can be repressed, but never quite dies.)

Plato had well thought to dispel this anguish by decree in his ideal state. The legislator, who aims at limiting the extravangant expenses met by relatives of the dead for their burial services, and at reducing the role played by the widespread cult of the dead in Greek religion,[5] decrees that the corpse is

37

just a "carcase of flesh," a mere image (*eidōlon*) of the dead, whereas the immortal soul is their real self.[6] Plato here clearly anticipates Aristotle's stance, who will dismiss the issue as just a matter of homonymy: "it is clear that a corpse is a human being in name only."[7] But Plato did not always share such an attitude. The corpse was still a phenomenon deserving to be saved in his early dialogues.

In the *Gorgias* he defines death, as almost verbatim in the *Phaedon*, the dissolution of the soul and the body from each other (*dialysis, tēs psychēs kai tou sōmatos, ap'allēloin*). But once they are disconnected, he goes on to argue, "each of them keeps its own condition very much as it was when the man was alive," so much so that even the medical treatments and the illnesses the body has suffered (*ta therapeumata kai ta pathēmata*) remain all manifest upon it.[8] Ulysses's recognition by his nurse in the *Odyssey* tellingly suggests how a scar might be used for the identification of an unknown body.[9] "Nocuments are Documents,"—as a seventeenth-century author captures the tragic pun *pathēmata mathēmata*[10]—in this case of an unaltered identity. Scars and other particular signs are still singled out by Origen as tokens of the continuing identity of a body over the span of its lifetime.[11] But the very endurance of the body after death is straightforwardly used as an argument *a fortiori* for the immortality of the soul in the *Phaedon*. Here, as in his dialogue with Charmides, Socrates has once again to recur to charms, but this time he applies them to dispel the anguish of "the child within us," who makes us fear that the soul might be blown away by the wind, once it goes out from the body (77E–78A). He invites his interlocutors to observe that

> when a man dies, the visible part of him, the body, which lies in the visible world and which we call the corpse, which is naturally subject to dissolution and decomposition, does not undergo these processes at once, but remains for a considerable time, and even for a very long time, if death takes place when the body is in good condition, and at a favourable time of the year. For when the body is shrunk and embalmed, as is done in Egypt, it remains almost entire for an incalculable time. And even if the body decay, some parts of it, such as the bones and sinews and all that, are, so to speak, indestructible.[12]

The—by necessity—only figurative immortality of the body nevertheless provides comforting evidence of the immortality of the soul: how on earth, Plato rhetorically asks, will the soul, which is noble, pure, and invisible, be straightaway scattered and destroyed when it departs from the body? The indefinite survival of the body in its mortal shape thus prefigures the eternal life of its alter ego. Such an opinion was also held by those Stoics Servius mentions in his commentary to the *Aeneid* (iii. 68), who maintained that the soul lasts as long as the body lasts, and praised the wisdom of the Egyptians in treating the corpses.[13] Yet to Plato incorruptibility remains at best an intimation of immortality.

Death is to him no longer the ultimate solution of continuity that comes to dissolve the transient unity of the body by absolutely loosening its limbs, as the Homeric formula for death decreed: *gyia lelynto*—nor the fatal solution of the bond tying the limbs together, as it still was for Euripides: *lelymai meleōn syndesma* (*Hipp.*, v. 199). Rather, it is a liberation of the soul from the body "as from fetters" (*ōsper ek desmōn ek tou sōmatos*; 67D), and in the attainment of this goal death and philosophy are objectively accomplices: so much so that "when you see a man troubled because he is going to die," Socrates argues, it is a sure sign "that he was not a lover of wisdom (*philosophos*) but a lover of the body (*philosōmatos*)."[14] The immortality of the soul demands the death of the body. And the body is not expected to rejoin the soul in the fruition of its eternal existence. The Apologists are the first who dared to bestow immortality on the human body, and made of resurrection, against the dualism of Neoplatonics and Gnostics, the true way of all flesh.[15]

Writing in the second century of our era, the philosopher Celsus disparagingly labels the Christians with the Platonic term of philosomatic sect (*philosōmaton genos*),[16] and evokes Heraclitus's authority in order to ridicule their "hope of worms" that the flesh may be granted an everlasting life: "as Heraclitus says, 'corpses ought to be thrown away as worse than dung'."[17] For Plotinus resurrection can only mean resurrection out of the body, certainly not with the body (*apo sōmatos, ou meta sōmatos, anastasis*; *Enn.* III.6.6). For Tertullian, on the other hand, "only that which has fallen can rise,"[18] hence the promise of the resurrection of the dead must refer to the part of man that is liable to fall, namely, *caducous*: the part Latin unequivocally calls *cadaver*, from *cadere*.[19] The soul, on the other hand, "has no name signifying falling," and rightly so, "because in its proper habit it does not collapse."[20] But Tertullian pursues even farther this exegesis *ad litteram*, which is perfectly consistent with his line of argument,—for a figurative reading would undermine his overall attempt to demonstrate the necessity of the resurrection of the flesh: "in fact, if all things are figures, what can that be of which they are figures? How can you hold out a mirror, if there is nowhere a face?"[21] Thus, by following step by step the account of the creation of Adam in *Genesis*, he reaches the conclusion that man is first and foremost flesh, for the name "man" was first used in reference to it:

> 'man' is first that which was formed, and afterwards is the whole man (*homo figmentum primo, dehinc totus.*) This submission I would offer, so that you may understand that whatsoever at all was provided and promised beforehand by God to man became a debt not

to the soul only but also to the flesh, if not by kindred of origin surely at least by prior possession of the name (*si non ex consortio generis certe vel ex privilegio nominis.*)[22]

The *restitutio in integrum* brought about by the final judgment will then reconstitute for eternity the texture of soul and flesh that makes up man, "carnis animaeque textura."[23] For the soul needs the flesh to live, as much as the flesh the soul: stunningly reversing Plato's dialectical identification of death and immortality, Tertullian literally reads its withdrawal from the flesh as a death for the soul, as well, and not just for the body: "to such a degree does the whole of the soul's living belong to the flesh, that to the soul to cease to live is exactly the same thing as to retire from the flesh."[24] For Tertullian, of course, the soul is also a body,[25] hence capable of being already subject to some forms of punishment while awaiting the *plenitudo resurrectionis*. But without the flesh the soul cannot achieve its perfection:

> for of its own it has no more than thought, will, desire, determination, while for accomplishment it awaits the activity of the flesh (*ad perficiendum autem operam carnis expectat.*) Likewise also for suffering it demands the alliance of the flesh, so as by means of it to be able as completely to suffer as without it it was unable completely to act.[26]

The sympathy of body and soul is thus the strongest argument in favour of their discontinuous, yet ultimately everlasting reunion. As Robert Klein put it, "according to the Church Fathers and theologians, the soul has no natural desire but to enjoy its body,"[27] the soul is moved, in Aquinas's terms, by an essential "appetitus naturalis ad corporis unionem."[28] The possibility that the corpse may eventually regain its former name marks therefore a decisive shift from the ancient conception of the body.

Dante has bent the nostalgia of the souls for their former bodies, their "desire of the dead bodies (*disio de' corpi morti*, *Par.* XIV, 63),"[29] to his own poetic service in the *Comedy*. The desire of embodiment is such, according to Statius's explanation (*Purg.* XXV, 79–108), that even in the interim period between death and resurrection the soul fabricates for itself a new body, made out of air, thanks to the still active power of its *virtù formativa*. Dante may thus justify, on this theological background, his stylistic device of providing the dead of a *corpo aereo*, capable of figuring the passions of the damned and the penitent (*secondo che ci affliggono i disiri/e li altri affetti, l'ombra si figura*, vv. 106–107),[30]—until their transfiguration will make any "creatural movement" (*movimenti umani*, *Par.* XXXIII, 37) obsolete in the aethereal transparency of Paradise.[31]

In the *Letter* he wrote "upon Occasion of the Death of his Intimate Friend," Sir Thomas Browne signals an "odd mortal Symptom," which he uncovered while caring for his dying friend but did not find "mention'd by Hyppocrates," namely, "to lose his own Face, and look like some of his near Relations." Browne interprets the uncanny sighting as due to an accelerated form of regression, as it were:

> for as from our beginning we run through variety of Looks, before we come to consistent and settled Faces; so before our End, by sick and languishing Alterations, we put on new Visages; and in our Retreat to Earth, may fall upon such Looks, which, from community of seminal Originals, were before latent in us.[32]

The effect of the approaching death is such that it forces us "to live backward," in a kind of vertiginous flesh-back, so to speak: the *morituri* do not just instantaneously review their previous existence in memory, they go through a continuum of metamorphosis. Once the will no longer controls the mien, death can willfully play with each individual feature and try unexperimented combinations of lines, virtual designs, "Caricatura Draughts."[33] Caricature is a proleptic figure of the work of death,—this is the first, and more encompassing, of the two interpretations Browne offers of the word he still spells in Italian, *caricatura*.[34]

It is also the first occurence in an English text of the word, which had only recently made its appearance in the Italian language, as well; more precisely in the atelier of the Carraccis, according to Giovanni Atanasio Mosini, who mentions the word for the first time in his preface to a collection of *Diverse figure* by Annibale (1646).[35] The word is used by Mosini to name the method of overloading or charging the features of a likeness, which makes out of it a "charged" portrait, a *ritratto carico*.[36] But already Baldinucci questions the Bolognese paternity of this "invenzione bizzarrissima," and reports that it was practiced in Florence since 1480.[37] There is indeed enough evidence to conclude that Leonardo was the first Renaissance artist to practice consciously and extensively the new genre as part of his overall "method of analysis and permutation."[38]

If nobody seriously draws into question the Renaissance genealogy,—unless one identifies *tout court* primitive art and caricature, as the first historian of the genre did[39]—no fully convincing hypothesis has been submitted to explain the "inexplicable delay," as Giorgio Agamben has observed, "attending the appearance of caricature in European culture."[40] According to Gombrich and Kris, who first remarked this "surprising" belatedness, "a free play with the representational image" can be experienced as "funny" only since the Renaissance, when "the taboo which had once forbidden the play with a person's likeness" is removed, and the belief in the magical power of images is for the first time dispelled.[41] If the hypothesis is psychologically seductive, it leaves out of sight the actual shapes the free play took; it lays down at most the conditions for the emergence of caricature, but does not help us envision the extravagant realities caricaturists created.

Browne's second interpretation of the term is certainly more suggestive. He relates straightforwardly the new technique of representation to physiognomy: "When Men's Faces are drawn with resemblance to some other Animals, the Italians call it to be drawn in Caricatura."[42] Undoubtedly, the engravings enclosed in Della Porta's treatise *De humana physiognomonia* are also specimens of the "discovery" that enabled the invention of caricature, namely, that "similarity is not essential to likeness."[43] By placing side by side on the same plate the figure of a man and that of an animal, Della Porta allowed the reader to remark with ease the resemblance of the two physiognomies on the whole, if not in detail, and to check at a glance, as it were, the validity of the so-called physiognomical syllogism, drawing conclusions on the moral qualities of a character from its proximity to an animal type [figs. 4–6].[44] We know from his biographers that Annibale intentionally pursued the genre, and also knowingly related it to physiognomy.[45] But this visual conceit had already been a source of entertainment to the ancient Greeks, if we trust Aristotle's testimony:

> humorists (*oi skoptontes*) often compare those whose strong point is no good looks in some cases with a fire-spouting-goat, in others with a butting ram; and there was a physiognomist who in his lectures used to show how all people's faces could be reduced to those of two or three animals, and very often he carried conviction with his audience.[46]

Animal analogies have indeed always been part of the vocabulary of art, as well as of ordinary language;[47] the beginnings of physiognomy may also be seen as an attempt at systematizing such a widely dispersed lore.[48] Through the anthropomorphic mirror of language,—the only harp Orpheus was ever in need of to tame the wild—we necessarily humanize animals. Adam, who names the animals according to a sign engraved on them by God, namely, according to their given character (the etymological meaning of the Greek word, according to Isidorus),[49] is not just the first onomaturgist, but also the first physiognomist: he reads their (internal) character through their (external) characters, and names them accordingly.[50] But how, then, do we name ourselves?

"Why when assigning their names to all the other creatures Adam did not assign one to himself" is the exegetical question Philo answers by denying to the earthly man[51] the faculty of self-mirroring, "just as the eye sees other objects but does not see itself."[52] Only the heavenly man, who was made "in the image, after the likeness of God,"[53] can know himself, and thereby His archetype, for he is the "offspring" (*gennēma*) of God, and not just a "moulded work (*plasma*) of the Artificer." He is the untarnished mirror of God.

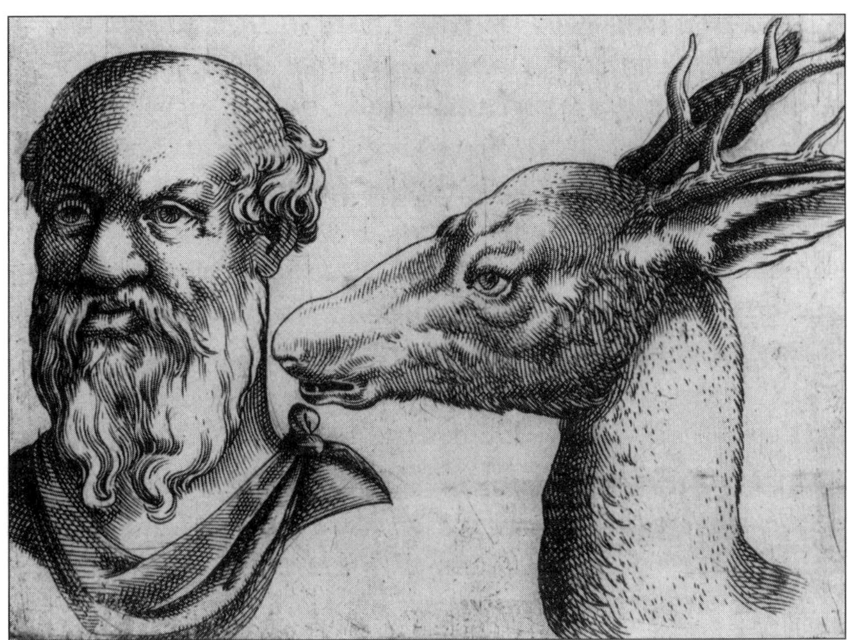

Fig. 4. Socrates-Stag, from Giambattista Della Porta, *De humana physiognomonia,* Vici Aequensi 1586. Special Collections Research Center, University of Chicago Library.

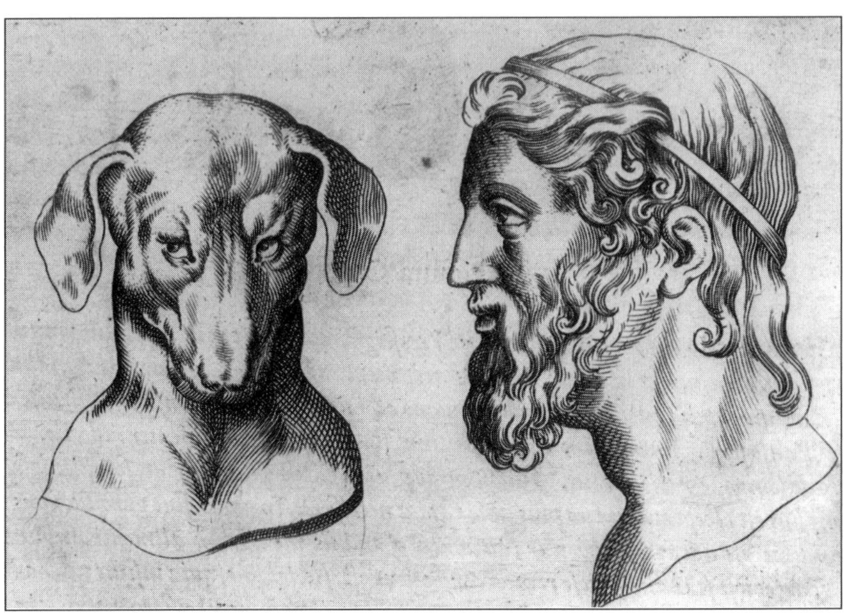

Fig. 5. Plato-Dog, from Giambattista Della Porta, *De humana physiognomonia,* Vici Aequensi 1586. Special Collections Research Center, University of Chicago Library.

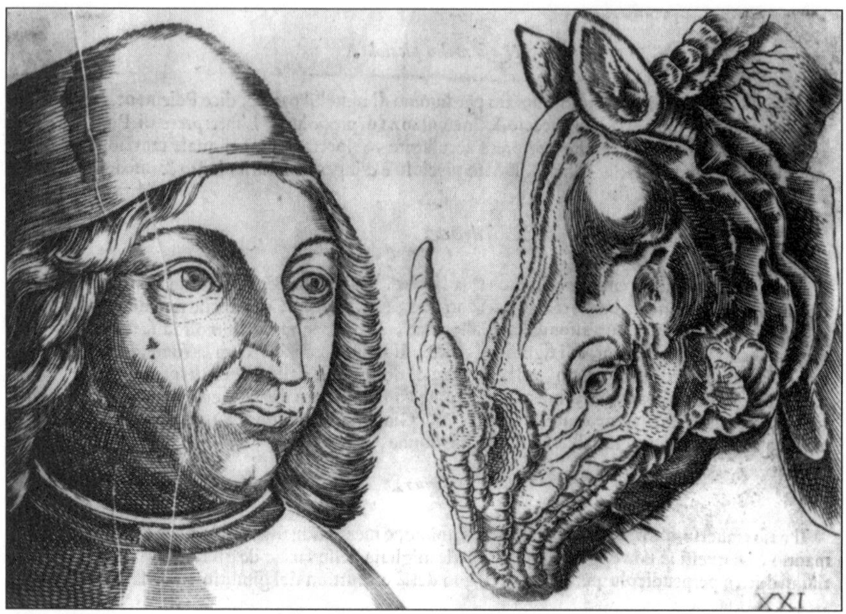

Fig. 6. Poliziano-Rhinoceros, from Giambattista Della Porta, *Della fisonomia dell' hvomo,* Napoli 1610. Special Collections Research Center, University of Chicago Library.

Whereas it is only after the slaying of Abel and the marking of Cain, when he is made aware of his mortality, that the earthly man can first read his name in the mirror of the other's face. The outline of physiognomy that Dante has sketched in a few lines of the *Divine Comedy* spells out the mark on Cain's brow:

> *Parean l'occhiaie anella sanza gemme:*
> *chi nel viso de li omini legge 'omo'*
> *ben avria quivi conosciuta l'emme.*

> Their eyes seemed like a ring that's lost its gems;
> and he who, in the face of man, would read
> OMO would here have recognized the M.[54]

The withdrawal of the flesh on the faces of the gluttons, condemned by *contrapasso* to suffer the pain of hunger, lets our generic name be more easily read on their features: "M being made by the two lines of their cheeks, arching over the Eyebrows to the nose, and their sunk eyes making O O which makes up *Omo.*"[55] [fig. 7] (Of course, the theory works best in Latin and Italian.) But it is clear enough that the sentence is cast not just upon those extenuated sinners, but on man as such: our common name is inscribed on

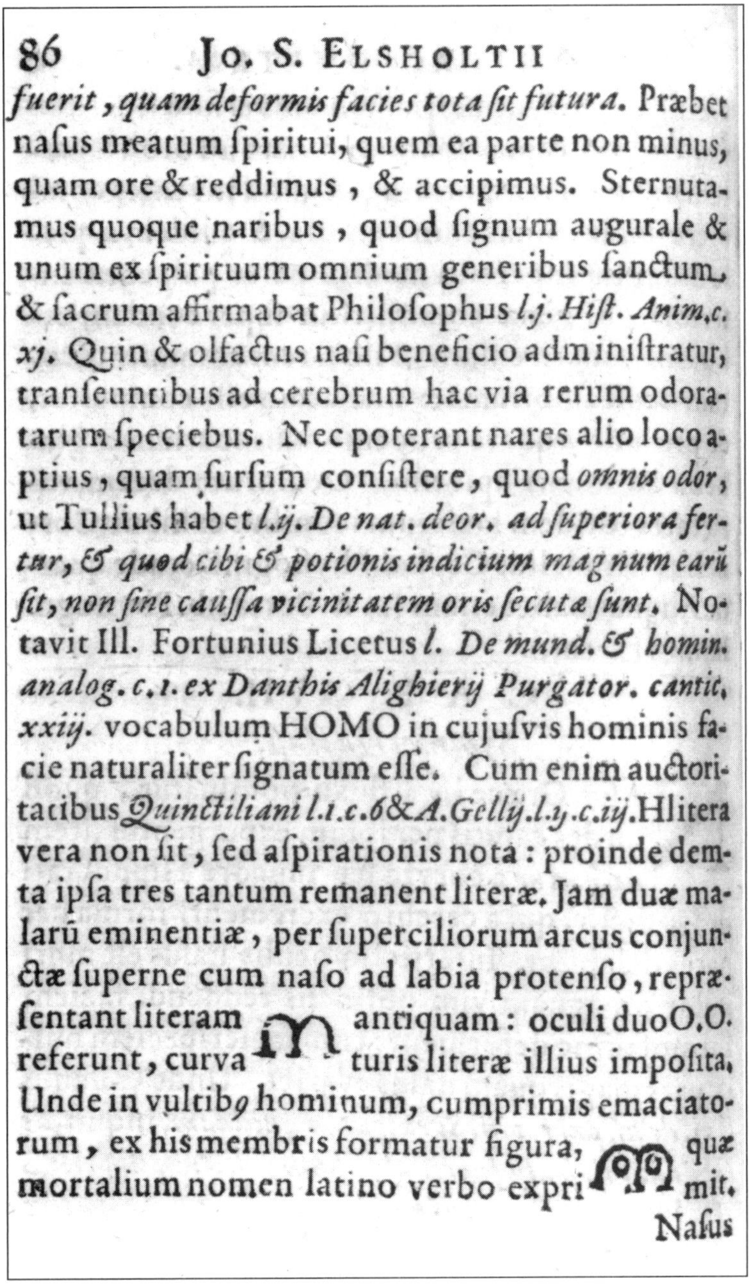

fuerit, quam deformis facies tota sit futura. Præbet
nasus meatum spiritui, quem ea parte non minus,
quam ore & reddimus, & accipimus. Sternuta-
mus quoque naribus, quod signum augurale &
unum ex spirituum omnium generibus sanctum,
& sacrum affirmabat Philosophus *l.j. Hist. Anim.c.*
xj. Quin & olfactus nasi beneficio administratur,
transeuntibus ad cerebrum hac via rerum odora-
tarum speciebus. Nec poterant nares alio loco a-
ptius, quam sursum consistere, quod *omnis odor,*
ut Tullius habet *l.ij. De nat. deor. ad superiora fer-*
tur, & quod cibi & potionis indicium magnum earū
sit, non sine caussa vicinitatem oris secutæ sunt. No-
tavit Ill. Fortunius Licetus *l. De mund. & homin.*
analog. c.1. ex Danthis Alighierij Purgator. cantic.
xxiij. vocabulum HOMO in cujusvis hominis facie naturaliter signatum esse. Cum enim auctori-
tacibus *Quinctiliani l.1.c.6 & A.Gellij.l.y.c.iij.* H litera
vera non sit, sed aspirationis nota: proinde dem-
ta ipsa tres tantum remanent literæ. Jam duæ ma-
larū eminentiæ, per superciliorum arcus conjun-
ctæ superne cum naso ad labia protenso, repræ-
sentant literam   [M]   antiquam: oculi duo O.O.
referunt, curva   turis literæ illius imposita.
Unde in vultibg hominum, cumprimis emaciato-
rum, ex his membris formatur figura, [symbol] quæ
mortalium nomen latino verbo expri [symbol] mit.

                                                           Nasus

Fig. 7. Johann Sigismund Elsholtz, *Anthropometria, sive, De mutua membrorum corporis humani proportione, & naevorum harmonia libellus,* Frankfurt an der Oder 1663. Special Collections Research Center, University of Chicago Library.

anybody's face,—on any mortal body, namely: "Aye, change if you can the moulding and stamp of the Divine coinage (*ei dynasai metaplatte kai metacharatte to theion nomisma*),"[56] thus Philo ridicules Cain's efforts to hide from His sight the character now indelebly printed, literally, in clear-cut letters, on the man's first-born's face.[57] It is the mark of our specific identity and at the same time the cipher of man's mortality, which death makes readable once and for all. "I never more lively beheld the starved Characters of Dante in any living Face," thus Browne describes the *facies* of his dying friend.[58] Caricature just lets characters surface.

One might also read Dante's verses as an epitome of Aristotelian physiognomy, which is throughout consistent with the tenets of the zoological taxonomy of the Stagirite. In it, type has decidedly the edge over idiosyncrasy. After all, Aristotle had forcefully argued that

> whenever the offspring fails to resemble its parents, we really have
> a sort of monstrosity (*teras*), since in these cases Nature has in a way
> (*tropon tina*) strayed from the generic type (*ek tou genous*);

but it is obvious that the prototype has decidedly the features of the father, for "the first beginning of this deviation is when a female is formed instead of a male."[59] In Aristotle's cosmos, alike in this to that of the Christian exegetes, the son is ideally the mirror-image of the father; reproduction, a mirroring. Any unlikeness, any shift from the type is strictly speaking a monster, or, in terms that were still unavailable to Aristotle, a caricature,[60] due to the resistance of matter to the inprint of the generic character. With the expulsion from Eden, it is clear that the mirror is tarnished. Still, Philo writes, Cain could only slain "that which shares Abel's name (*to homonymon tou Abel*)," "the impression stamped to resemble him (*ton apeikonisthenta typon*)," certainly not "the original (*to archetypon*)."[61] The Son of God and the son of man are only homonymically related. The archetype remains untainted by the degeneration of mankind, He is still "the Image of the Father, the same stamp upon the same metall," as Donne rewords Philo's numismatic analogy, whereas the sinful man is "a peece of rusty copper," in which the same Image of God had been originally stamped but has been since "defaced and worn."[62] The history of mankind is properly the history of man's decline from his type. Man's history is a story of decadence and degeneration, from the man first fashioned, who was clearly "the bloom (*akmē*)" of our race, downwards,[63] a regression from the face of God into that vast "region of unlikeness" (*regio dissimilitudinis*), where Augustine—thanks to his reading of "certain books of the Platonists"—realizes he has lost himself.[64] How can man, who is "from head to foot contained in space" (*a capite usque ad pedes in loco*), be an image of God? [65] On the face of such an incongruity Augustine embraces the spiritual reading of the Scriptures advocated by Ambrose under the aegis of Paul's authority (2 *Cor.* III.6): "The letter killeth, but the spirit giveth life."[66]

On the other hand, the seemingly literal, albeit apocryphal interpretation of the opening line of Genesis, as meaning "In the beginning God made for himself a son," which Tertullian records in his treatise *Against Praxeas* (V.1),[67] undoubtedly promises a more intriguing account of God's creation. "Who is the Father?" (*Quis enim pater?*)[68] is the unorthodox but completely legitimate question Tertullian poses from his literalistic stance. "Father" is a relative term, hence it necessarily entails its counterpart.[69] "How can you hold out a mirror, if there is nowhere a face?" The mirror is the Son, the face is the Father; but a mirror has no face of its own. Hence the Father is "the Son's face" (*facies filii pater*), [70] but not vice versa: the relationship is not symmetrical, as the mirror-imagery might suggest. We bless the Son in the Name of the Father, but we can know the Father only in the Face of the Son. The Son is the visibility of God.[71] He is also a persona of the Father, in the trinitary sense of the word, but the relationship between the two is by definition closer than impersonation, it is—in *the* proper sense of the word—filiation.

Be it literally or spiritually interpreted, filiation will remain throughout the history of Western culture the model of a likeness that imitation can at best emulate, but never equal.[72] Yet, as opposed to the relationship of the Son of God to His Father, the relation of similarity between man and his Creator must take time into account: it is a gradual approximation, a becoming-alike, and not an isomorphism; an apotheosis,[73] and not an anthropomorphism. As Tertullian writes in his treatise *On Baptism*, "imago in effigie, similitudo in aeternitate censetur," [74] the adequacy of a likeness is evaluated by comparing the original to the image, but likeness itself can be tested only by the standard of eternity. In Kierkegaard's secular terms, this means that, to the son, the father "is like a mirror in which he beholds himself in the time to come."[75] [fig. 8] Yet outgrowth is no simile.

"Growth and form" was an hendiadys in ancient biology; the obsolescence of teleology made of it an oxymoron.[76] Goethe could still define the organic a "minted form that lives and living grows" (*geprägte Form, die lebend sich entwickelt*).[77] But such a definition appeared already to Simmel a contradiction *in terminis*. How can a form develop? How can a character grow?

> How can that which has once been "minted" still "develop" itself, does "character" have a meaning at all, if it does not persist over a certain span of time, but is rather a ceaseless transformation?[78]

This continuous change is what Simmel, using with noticeable irony the Goethean term *par excellence*, calls the *Urphänomen* of human life.[79] Life is the shattering of form. The inscribed character claims for itself eternity, but time reads it as caricature. Putting it otherwise, caricature is nature *before* or *after*, but not *at* its acme. Goethe's definition of life thus reads like the formula of caricature. Caricature is "the foundering of form against life."[80]

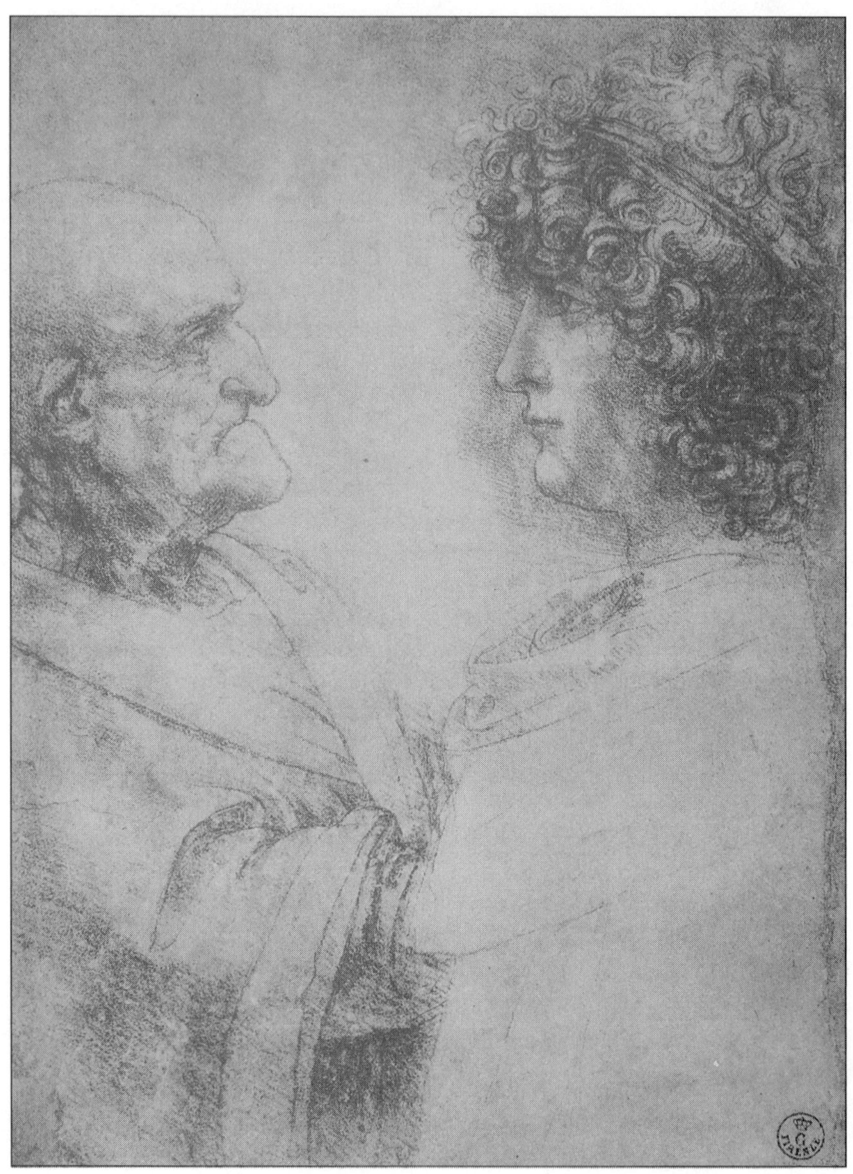

Fig. 8. Leonardo da Vinci, *Studio per volto: testa di vecchio e di giovane*. Drawing, Gabinetto dei Disegni e delle Stampe, Museo degli Uffizi, Firenze. Photo: Biblioteca della Stamperia d'Arte-Fratelli Alinari, Firenze.

In justifying his *passatempi* Annibale Carracci claims to have followed the example of Raphael, who gathered features from different individuals in order to reach his ideal of beauty; [81] and describes his own aim as the equally ambitious one of achieving a "perfect deformity" (*perfetta deformità*).[82] An oxymoron in the Aristotelian universe, "perfect deformity" is the intentional parody of the "perfect proportion" Vincenzio Danti, writing under the influence of Michelangelo, had vowed to codify in his treatise *Delle perfette proporzioni* (1567).[83] The caricaturist thus mimics the idealizing strain of high art on both the theoretical and the practical level, yet he too lightheartedly misrepresents the professed intentions of the idealizing artist. In his famous letter to Castiglione Raphael indeed precisely voices his dissatisfaction with the method of *electio*, or "selective imitation," and claims the crucial role of the idea in the creative process.[84]

In embracing an already outmoded view of the creation of beauty,[85] and making of it the model *e contrario* for his quest of the "beauty of deformity" (*la bellezza della deformità*), Annibale betrays the archaic bent of the ambiguous *genre* he masters.[86] The caricaturist certainly does not feel obliged to fulfil the promise of wholeness the idea heralds. The methods of *electio* and caricature objectively converge in that they both make of the isolated feature the focus of attention. The *perfetta deformità* to which Annibale aspires is achieved by "aggravating" features and "distorting" proportions;[87] his intention is directed toward the individual features, which he magnifies and then reassembles. He can thus fully exploit the infinite combinatory possibilities therein contained, whereas the idealizing artist remains confined to the repetition of the type.

Yet unquestionably the legitimation of both modes of representation (idealizing portraiture no less than deforming caricature) was only possible once the detachment of the image of man from that of God had made of the dignity of man a value in itself, and not a matter of coinage. In his *Oratio* (1486) Giovanni Pico della Mirandola had written that God, "summus Pater" but also "architectus,"[88] created man as "a work of indeterminate image" (*indiscretae opus imaginis*) and placed him in the middle of the world. God's apostrophe to his creature stresses his facelessness:

> Neither a fixed abode nor a form that is thine alone (*propriam faciem*) nor any function peculiar to thyself have we given thee, Adam, to the end that according to thy longing and according to thy judgement thou mayest have and possess what abode, what form (*faciem*), and what functions thou thyself shalt desire [. . .] We have made thee neither of heaven nor of earth, neither mortal nor immortal, so that with freedom of choice and with honor, as though the maker and molder of thyself, thou mayest fashion thyself in whatever shape

thou shalt prefer (*ut tui ipsius quasi arbitrarius honorariusque plastes et fictor, in quam malueris tute formam effingas*).[89]

The boldness of such a statement is apparent if compared with the mirror-cabinet of the Neoplatonic-Christian world, and its methodological importance for the figurative arts can be hardly overrated. In a sense, the traditional God of the biblical exegesis, who creates man in *His* own image and likeness, becomes now the prototype of the bad artist.[90] Likeness is no longer anchored in an ontological order, painting and sculpture are liberated from any repetition-compulsion. Writing only a few years earlier his treatise *On Sculpture*, Alberti is still very cautious (*si recte interpretor*) in arguing that there are two different ways of achieving likeness, one aiming at the type, the other at the individual.[91] Even more indicative of the author's still unresolved mind is the aporetical discussion of the nature of "likeness" (*similitudo*) that precedes these considerations: on the one hand we see, Alberti writes, that nature "is accustomed to observe in any creature that each is like the rest of its kind (*ut eorum quodque sui generis quibusque persimillimum sit*);" and yet, on the other hand, "as they say (*ut aiunt*), no voice or nose or similar part resembles any other among all the rest of the people."[92] We may safely recognize the latter remark as belonging to the credo of the Renaissance physiognomist, who engages in the "hunting" (*venatio*) of individual differences[93] guided more by discerning judgment than by analogical wit.

The artist, who oscillates between portrait and ideal, is thus caught in a dilemma that reflects the very ambivalence of nature between the generic and the individual. This dilemma spans the whole of Renaissance art.[94] But certainly, among his contemporaries, no other artist or theoretician was more aware of the dangers therein involved than Leonardo; no other made of the declension from the type such a central concern of his own artistic endeavors, at least on a theoretical level. Again and again[95] Leonardo warns the artist against the unconscious repetition of the same that would mar his creations, even though he was himself a victim of such a compulsion, and *pour cause*, as Freud has shown.[96]

Beyond scattered remarks throughout the general precepts in his treatise on painting, Leonardo devotes a special paragraph to what he condemns as "the greatest fault of painters"—their compulsion to repeat.[97] Such a fault is virtually ineradicable, since the same soul that has built up the body where it dwells must keep operating in the same fashion in any of its further creations:

the soul which rules and directs each body is really that which forms our judgement before it is our own judgement. Thus it has developed the whole shape of a man, as it has deemed to be best with long, or short, or flat nose, and definitely assigned his height and shape. This judgment is so powerful that it moves the painter's arm and makes

> him copy himself (*fagli replicare se medesimo*), since it seems to
> that soul that this is the true way to construct a man (*il uero modo
> di figurare l'homo*), and whoever does not so, commits an error.[98]

The compulsion of the soul Leonardo describes in these terms is of the same kind as the *conatus* of an element toward its natural *locus* in the ancient cosmos, it is a natural inclination the soul cannot but comply with, like the souls of Dante's penitents, which refashion their bodies out of air as soon as they are assigned to their temporary destination.[99] Leonardo's "judgment" is therefore to be regarded as a power rather akin to Goethe's *daimōn* than to Freud's unconscious,[100]—if one must needs look for a metempsychosis of the concept. For the word *giudizio* must not be taken here in a strictly psychological,[101] but rather in a "wholly naturalistic" sense, as referring to a natural inclination that cannot be persuaded.[102] Artists cannot refrain from repeating themselves, and the theoretician can only record such a feature of natural history, as it were.

A suitable *trait d'union* between Leonardo's and Goethe's insights is provided by Shaftesbury's enthusiastic aphorism: "The *characteristic* [. . .] is all in all."[103] *Daimōn* is, according to Goethe's own gloss to the *Urworte*, precisely "the characteristic (*das Charakteristische*), through which each one distinguishes oneself from everybody else in spite of an however great similarity;" while, at the same time, this identifying trait vouches for "the immutability of the individual" (*die Unveränderlichkeit des Individuums*). In a handwritten draft of the *Urworte* Goethe straightforwardly interprets the Greek word as "Individualität, Charakter."[104] It is plain that the word "character" must be taken here in a stronger sense than the one we ordinarily associate with it: as meaning not just firmness, but rather inalterability of character. Because of such an intransigence, of such a single-mindedness, so to speak, the ancient concept of character, which Goethe properly translates with *daimo͞n* and not with *e͞thos*,[105] must be assigned to the sphere of nature, and not to that of ethics.[106] The moral of the fable, as Shaftesbury puts it, can only be an ostensive gesture: "such a one he is! Such he is—*Sic, Crito est hic!* This is the creature!"[107]

Like God, the artist cannot but express his own character in the character of his work. Or, to put it otherwise: he cannot dissimulate the presence of his own character in his own work. Still, such a necessity turns into a virtue if the artist is endowed by his soul of a grace "in excess of measure," as Vasari demands;[108] counterfeits (*contrafazioni*),[109] on the other hand, are the result when the artist lacks this natural standard of beauty. In order to avoid that one's own body become a habit, the artist has then no other choice but to resort to the method of *electio* and to the standard of common judgment.

Yet these *temperaturae*[110] of one's own misjudgment are legitimate insofar as the very variety of Nature suggests that there is no universally valid rule of judgment: "Therefore you, imitator of such nature, note and be attentive to the variety of features" (*Adunque tu, imitatore di tal natura, guarda et attende alla varietà de' lineamenti*).[111] *Quot capita, tot sententiae*, nobody can make of, literally, one's own head the standard of judgment, if not by (a purely quantitative) approximation. For Nature has only set rules for the quantities of the individual features, not for their qualities:

> If nature had fixed a single rule for the qualities of the features (*membra*), all the faces (*visi*) of men would resemble each other in such a way that it would not be possible to distinguish one from the other. But she has varied the five parts of the face (*i cinque membri del volto*)[112] in such a way that, although she has made an almost universal rule for their sizes (*grandezza*), she has not observed it in the qualities in such a way that it is possible to recognise clearly one person from another.[113]

In other words, there is a limited range of possible variations in the size, an indefinite one[114] in the shape of the features of a face, and this accounts for both the likeness and the unlikeness to its kindred each human face is bent to display. Still, for the sake of instruction, given "that it is impossible to memorize all the aspects and changes of the parts of the body," the number of variations of each feature can be reduced to a limited figure: the nose, for example, can show ten possible shapes in profile, eleven in full face [fig. 9]; and so forth for the other members. These classifications are indispensable for all practical purposes, and especially when the painter has to work out of memory, or *a forza di memoria*. In this case the painter must be able to review in his mind the impression made upon him by a face no longer present.[115] Leonardo puts forward a method that enables the painter to recompose the face out of its *disjecta membra*:

> when you have to draw a face from memory (*quando hai a fare un uolto a mente*) take with you a little book wherein are noted down similar features (*factioni*), and when you have glanced at the face of the person you are to portray, look then at the parts, which nose, or mouth is like his, and make a little mark to recognize it, and then at home put it together (*poi a casa mettilo insieme*).[116]

It remains untold, however, whether this synthesis will be able to recompose the analyzed whole in such a way as to convey its individuality, the quality that Italian painters had been calling—for a century at least before Leonardo— the *air* of a face.[117]

In one of the rare lyrical moments of his notebooks, which keep otherwise a consistent tone of impassive scientific rigor, Leonardo invites the student to go out "in the streets, as evening falls, and when the wheather is

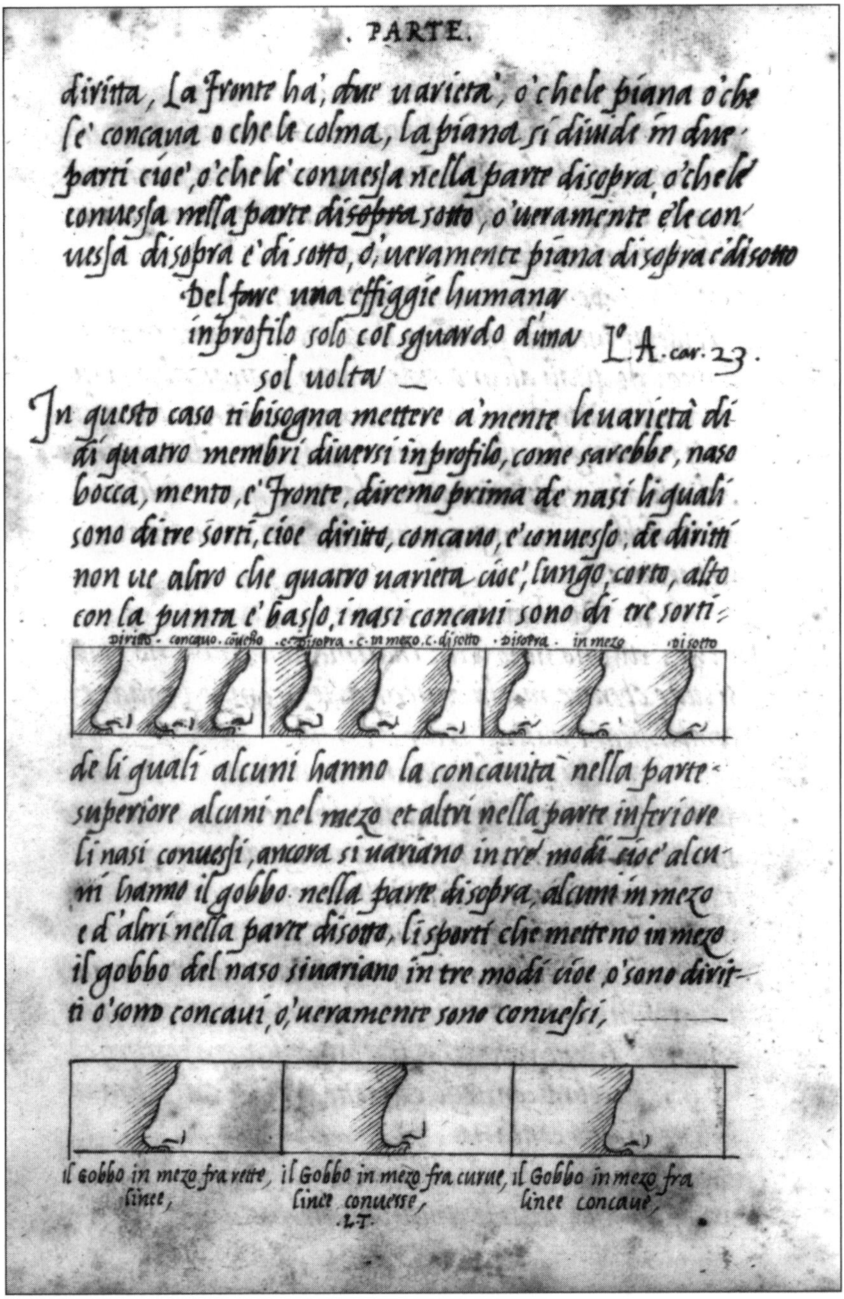

Fig. 9. Leonardo da Vinci, *Libro di Pittura*, Cod. Urb. Lat. 1270, f. 108v. Photo: Biblioteca Apostolica Vaticana.

dull," for these are the most propitious conditions of air under which to observe the faces of men and women: "what grace and delicacy you may perceive in them!"[118] Only when the surrounding air lends its favor[119] to a face can the face, in turn, display grace and delicacy.[120] On the other hand, the only thing the zelous *membrificatore*[121] seemingly can bring home, as the previous annotation suggests, is a series of marks on his notebook. The members can then be put together at leisure, but the "grace and delicacy" has probably vanished in the process, as it was dependent from the concomitant occurrence of the ideal atmospheric circumstances, what Leonardo calls "perfetta aria."[122] The air affects the face in ways that "remain no longer in their subjects."[123] Therefore, these "fugitive or transient beauties"[124] can only be caught "to the moment," as it were, but not remembered. Only the features of a face can be remembered. A portrait from memory will necessarily miss the air of a face.

Remarkably, monstrous faces are excluded from Leonardo's mnemonics, "since these are remembered without effort." Probably because monstrous are those faces in which one single member, grown out of measure, eclipses all the others and engrosses entirely our attention [fig. 10]. A single feature cannot be remembered, but is nonetheless unforgettable. Walter Benjamin writes that, in Molière's comedies, "character develops like a sun, in the brilliance of its single trait, which allows no other to remain visible in its proximity, but outshines it."[125] In a world devoid of character, single-mindedness inevitably becomes an object of ridicule. The man of character turns into a singular man, an original (*ein Sonderling*), who may at will be drawn in caricature, as Kant laments.[126] Caricature is, also, the modernized spelling of character.

After decreeing its obsolescence, Kant suggested that we interpret immortality as "the necessary permanence of the personality."[127] A logical, though fateful, consequence of such a translation of immortality is the identification of "face" with "persona." The face no longer carries around the stamp of a character, but is the badge of a personality. After the demise of religion, the task of philosophy has precisely become, according to Shaftesbury,

> to teach us *our-selves*, keep us the *self-same* Persons, [. . .] as to make us comprehensible to our-selves, and knowable by other Features than those of a bare Countenance. For 'tis not certainly by virtue of our Face merely, that we are *our-selves*.[128]

Thus, since the Enlightenment, we can only register gains in personality,[129] to the point that the build-up of the personality has now become the answer to the shrinking of the aura.[130] Yet, as Kant well knew, if one takes away the mask, what remains is a thing, rather than a face: "*eripitur persona manet res*."[131] The danger is that Hegel's reduction *ad absurdum* of physiognomy might be taken as a self-evident truth, and his ironic assimilation of the spirit to a bone be considered as an immediate token of redemption.[132] Yet a bone without spirit is as hollow as a bone without marrow.

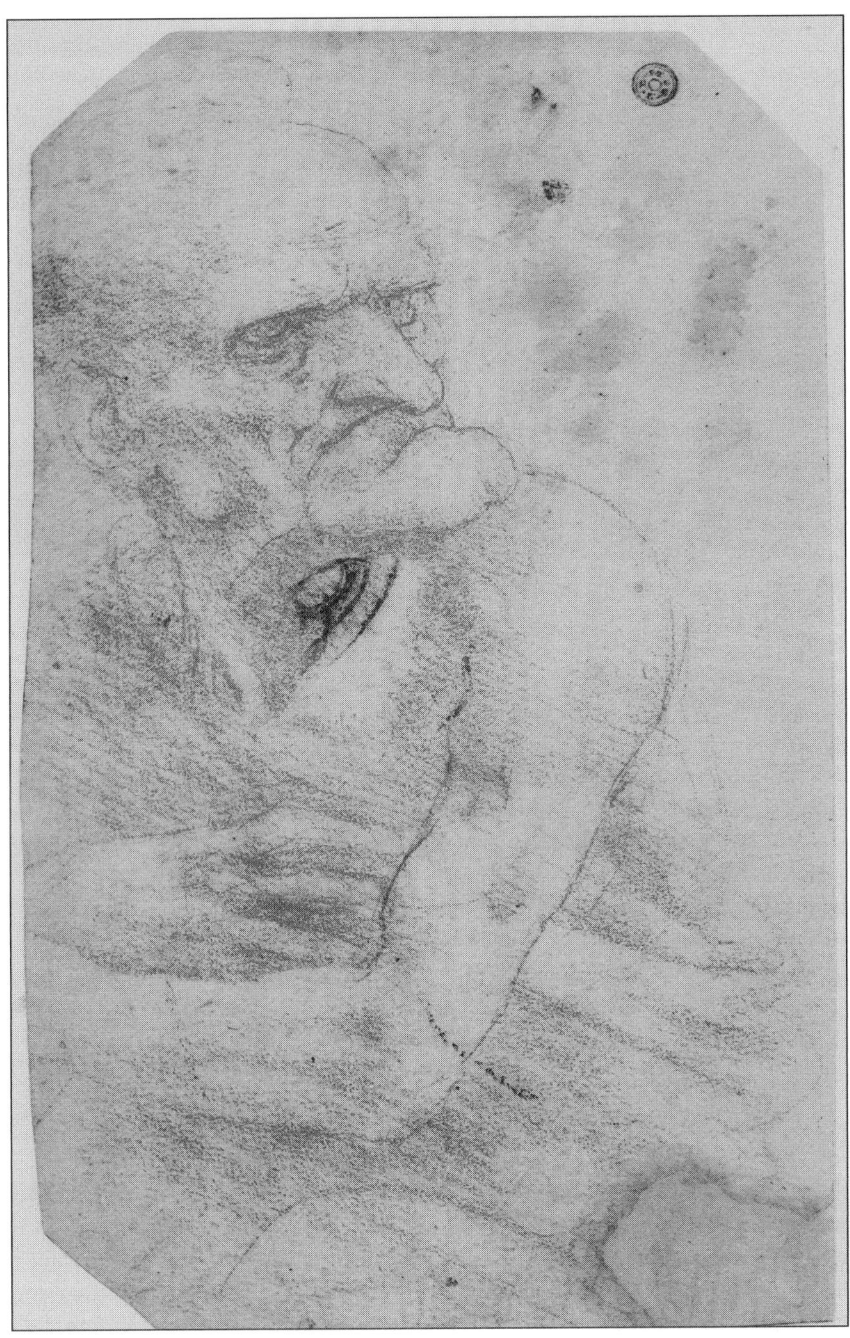

Fig. 10. Leonardo da Vinci, *Caricatura di Giuda, studio per il Cenacolo*. Drawing, Gabinetto delle Stampe, Galleria d'Arte Antica, Palazzo Corsini, Roma. Photo: Biblioteca della Stamperia d'Arte-Fratelli Alinari, Firenze.

The still most practical way of establishing the death of a body, and the only one *de facto* available until the end of the eighteenth century—with the possible exception of a feather, as in the final scene of *King Lear*[133]—is by approaching a mirror to its lips: when the breath no longer mists the polished surface of the mirror, the transformation of the body into a corpse may be certified. The mirrored image of the face of the dead, which the breath has no longer been able to blur away, offers to the sight a still life, Kent's "promised end," or "the image of that horror," if we prefer Edgar's description.[134] Another problem confronting Renaissance theorists of art, and as challenging as that posed by the representation of the "air" of a human face, had been indeed that of the representation of a dead body—one most difficult to solve (*quod quidem difficillimum est*), according to Alberti, because of the impossibility of representing absolute absence of motion.[135]

Anatomical illustration is another late offspring of European culture, as the taboo that lastingly protected the human body from autopsy, apart from the Alexandrian parenthesis, was definitively broken in the Renaissance. The problem of representing a dead body poses itself with particular urgence to the anatomist, who must analyze and describe the functions of a living organism on the basis of a dissection that can never replace vivisection. To bypass such an obstacle, and for obvious devotional implications, the anatomical iconography *ab ovo* resorted to a fictitious resurrection of the flesh, setting the stage for a prefiguration of the last judgment. Vesalius's tables thus present us with a parade of skeletons and *écorchés*, depending whether the bony or the muscular structure of the human body is illustrated. The poses and the attitudes are those of a living body; what is missing is the envelope that, in a living body, hides from sight its interior, namely, the skin. In an engraving from Juan de Valverde's treatise of anatomy (1556) [fig. 11], an *écorché* brandishes like a spoil his cast-off skin, hanging like the dead body of which Alberti laments the difficulty of representation: its members "hang loose; hands, fingers, neck, all drop inertly down" (*omnia pendent, manus, digiti, cervix, omnia languida decidunt*).[136] The engraving is the work of the Spanish painter Gaspar Becerra, who had collaborated with Michelangelo to the frescoes of the Cappella Sistina.[137] Becerra is plainly alluding to the self-portrait Michelangelo had inserted in the Giudizio Universale [figs. 12–13] under the guise of Saint Bartholomew, the martyr who had been flayed alive, as Marsias by Apollo, two favorite subjects of later Mannerist painting. By taking apart the body and its envelope, Michelangelo ingeniously solved the problem of prefiguring his own aspect *in articulo mortis*, and produced what one might regard as an ideal caricature.[138] Such is the legacy the Renaissance at its apex hands down to its epigones. Thereafter the flesh, so to speak, will retreat and let surface the natural *ritratto*[139] of the face: the skull, which will dominate the Baroque imagery. Yet a skull no longer provides comforting evidence for the recognition of a face, be it even a face whose lips one has kissed one knows not how oft.[140] Out of the air there is no "favor"[141] to a face.

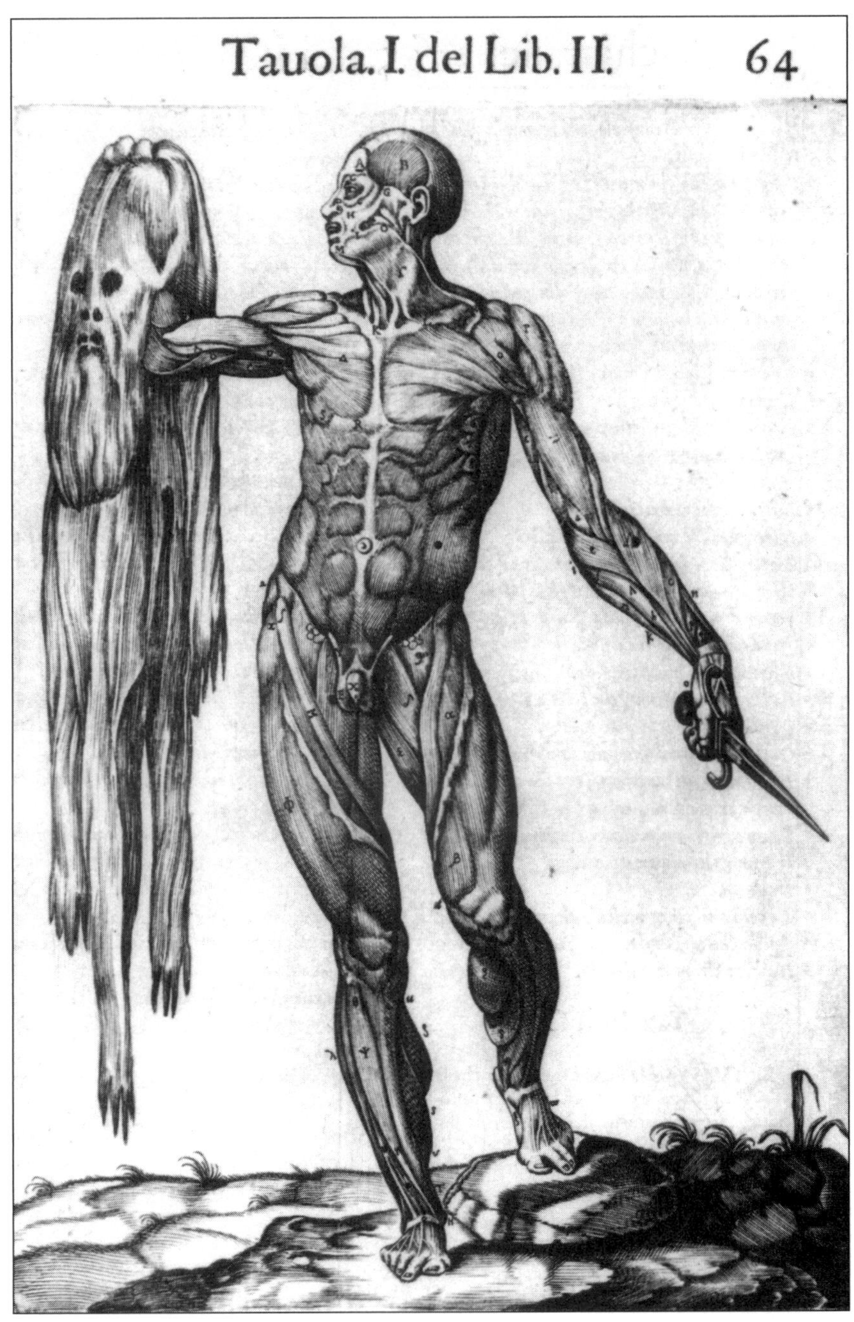

Fig. 11. Juan de Valverde, *Anatomia del corpo humano,* Roma 1558. Special Collections Research Center, University of Chicago Library.

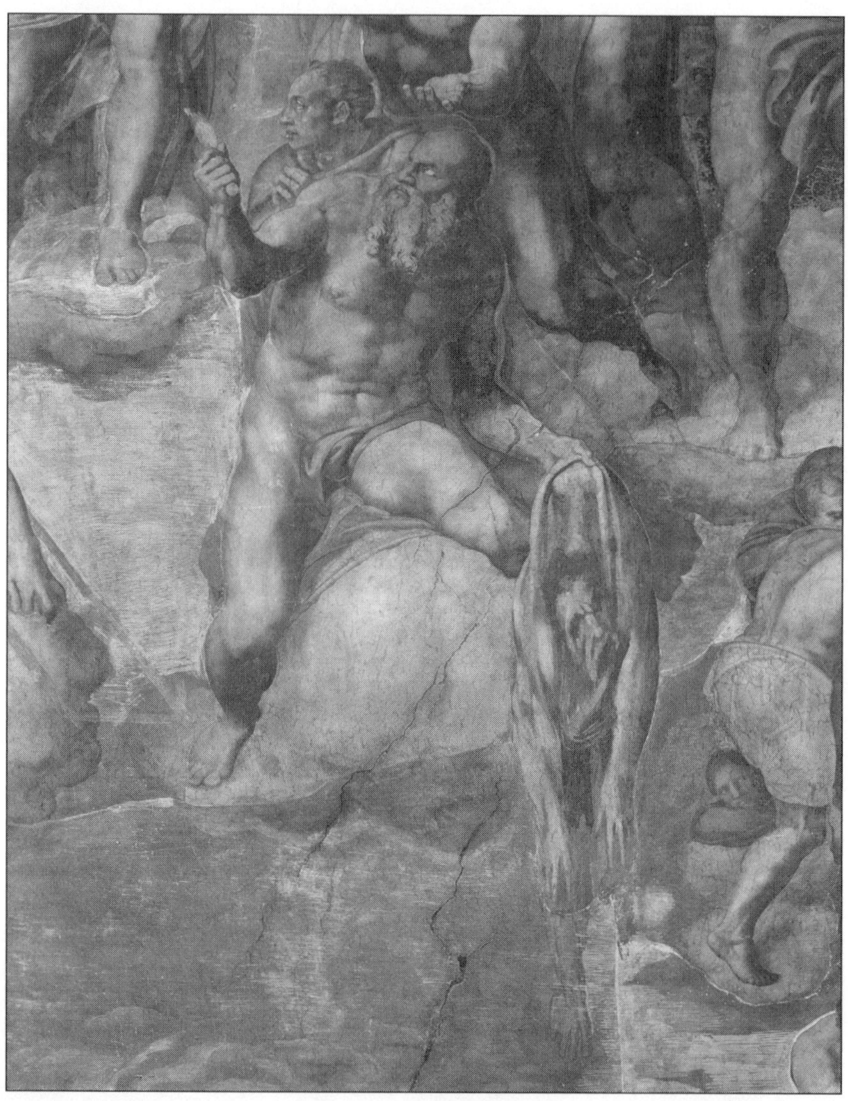

Fig. 12. Michelangelo, *Saint Bartholomew,* detail of Last Judgement, Sistine Chapel, Vatican Palace, Vatican State. Photo: Scala/Art Resource, New York.

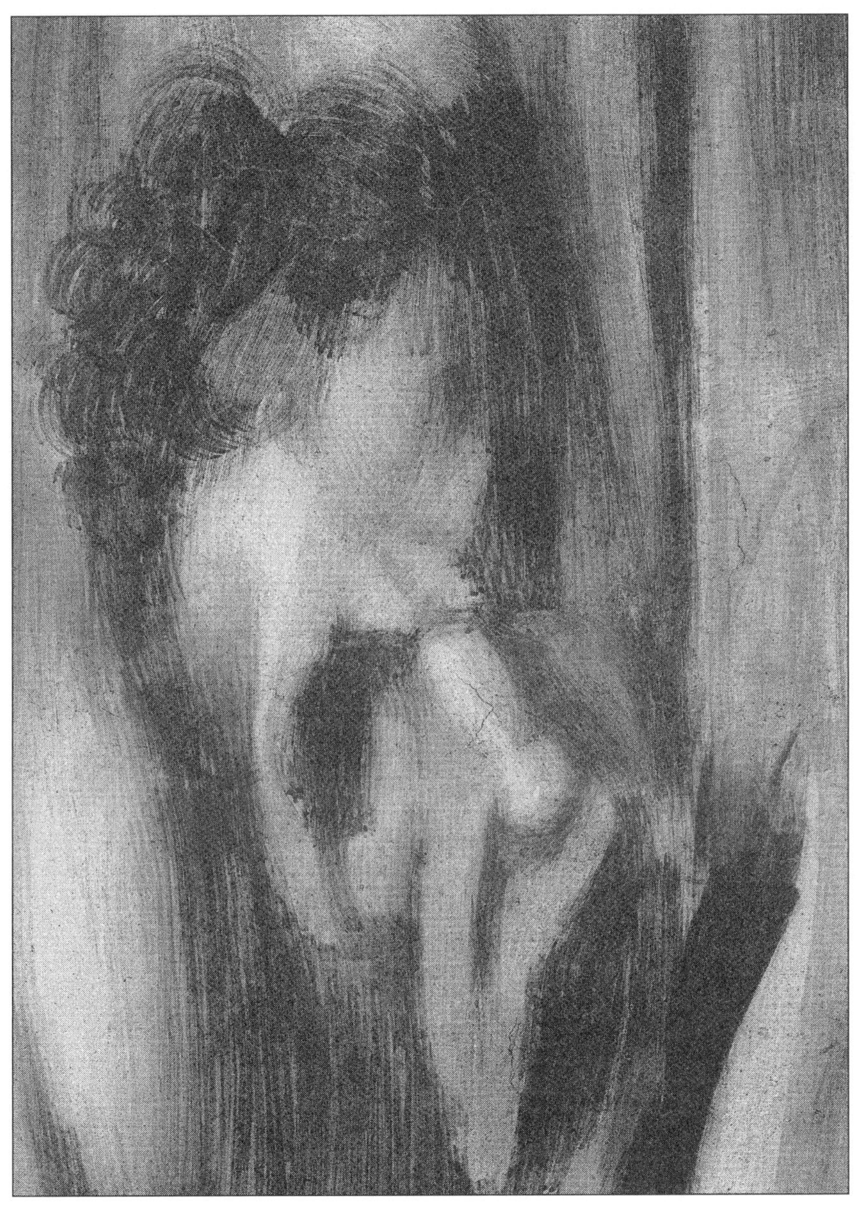

Fig. 13. Michelangelo's self-portrait. Detail of fig. 12.

Chapter 3

Dysōpia and Discountenance

HOLOFERNES I will not be put out of countenance
BIRON Because thou hast no face.

Love's Labour's Lost V.2

"Vitiosa verecundia" is the paraphrasis Erasmus opted for as a translation of Plutarch's *dysōpia*.[1] He could not rely on any standard equivalent, as even Cicero, the foremost advocate of the translatability of Greek into Latin, uses twice the Greek term in his letters without rendering it into Latin.[2] However, differently from other Greek words I have been discussing,—such as *charactēr*, *symmetria*, or *harmonia*—*dysōpia* has not been grafted onto the stem of the modern European languages. The choice of most translators, following Erasmus, has been to add a pejorative epithet to a word meaning "shyness" or "shame." Philemon Holland, the first English translator of Plutarch's *Moralia*, qualifies "bashfulness" with the adjectives "unseemly and naughty."[3] The Latin translation by Xylander, which has become standard, is "vitiosus pudor;"[4] Amyot translates *dysōpia* with "fausse honte"[5] and a 1549 Italian translation by Antonio Massa has "quella erubescentia, che è vitiosa & dannosa."[6] The first German translation, published in 1617, resorts to an even longer periphrasis: "die zuviel unziemliche bäwerische Schamhafftigkeit."[7] Kaltwasser validates the abbreviated form "Bauernscham" in a footnote to his classical translation of the *Moralia*,[8] but he prefers to it "die falsche Schamhaftigkeit," as, before him, Nüscheler.[9]

In spite of their ingenuity, however, all these translations are erroneous, at least insofar as they suggest *dysōpia* to be a vice, or a false shame, as opposed to a true, virtuous one. In Plutarch's view, as well as in that of his main source, the Aristotelian ethics, neither is *dysōpia* a vice, nor is shame a virtue. They are rather physiognomical signs of a certain moral disposition of the mind, "indicia et imagines virtutis" rather than virtues themselves, as Francesco Piccolomini glosses Aristotle in his *Universa philosophia de moribus*.[10] They are signs of a fundamentally healthy moral constitution in their subjects, in the same way in which "wild and unproductive plants" are the "signs of a soil not unfertile, but generous and rich:"

> So too with the affections of the mind (*pathē psychēs*): some that are bad are nevertheless the outgrowths, as it were, of an excellent nature well able to respond to the cultivation of reason.[11]

Such is the case of *dysōpia*, which Plutarch defines as an excess of shame (or even an "hyperbolic" shame: *hyperbolē tou aischynesthai*, 528 E) "that compels us to grant an unjustified request."[12] It has been convincingly argued that the word was first introduced in the philosophical terminology by the Stoics,[13] but Plutarch's definition is more directly indebted to the Aristotelian taxonomy. In the *Nicomachean Ethics* (1128b10–30) Aristotle defines shame (*aidōs*) as "fear of ill-repute," and declares it a passion, rather than a habit, because, as the blushing of the shameful openly reveals, it is "in some way connected with the body, and this is held to belong rather to a passion (*pathous*) than to a habit (*hexeōs*)." Aristotle's definition of shame thus already entails its appearance.[14] Plutarch, too, regards *dysōpia* as a passion, and its physical symptoms as so eye-catching that the etymon of the word cannot but reflect them: "Hence the name, the face being somehow (*tropon tina*) involved in the embarassment and discomposure of the mind (*tou prosōpou tēi psychēi syndiatrepomenou kai synexatonountos*)" (528 E). The modality, or the tropology, of such a discomposure, however, is not at all unambiguous.

Later English translators have generally adopted Holland's choice, with the exception of the editors of Plutarch's text for the Loeb Classical Library, who have proposed instead "compliancy."[15] The term "compliancy," however, fails to convey the physiognomical implications of this moral category, which come to the foreground in Plutarch's discussion, as we have already seen. At the end of the chapter I will propose my own translation of the Greek term. I prefer it for reasons that, I hope, will become apparent as my argument unfolds.

Plutarch's treatise is meant in part as an advisory against the threat of flatterers, a species that is certainly not extinct, but must have been particularly annoying in the Greek-Roman world.[16] Theophrastus includes a profile of the flatterer among his *Characters*, and Plutarch himself devotes a special treatise to the problem of distinguishing the flatterer from the friend, probably inspired by a writing *On Flattery* by the Peripathetic Ariston of Ceos.[17] In Theophrastus's sketch, one of the slick ways in which the flatterer pursues his patron's favor is by praising the resemblance of his portrait (*tēn eikona omoian einai*).[18] And Plutarch warns against the inclination to give in to the flattery of being portrayed as a symptom of *dysōpia*: "let us break ourselves from using either a barber to trim us, or *a painter to draw our picture*, for to satisfie the appetite of our foolish shamefacednesse."[19]

Certainly one could not put the blame of such a weakness on Plotinus. Porphyry starts his biography by consciously distancing himself from the conventions of the genre, which would demand a literary portrait to be prefaced to any prosopographical work. He excuses himself for such an infraction of *decorum* by recording that his master was so ashamed—*not* abashed—of being in a body he stubbornly refused to sit for a portrait his disciples had requested of him. In Porphyry's account, Plotinus justifies his objections to portraiture on the basis of a standard Platonic argument:

> Is it not enough to have to carry the image (*eidōlon*) in which nature has encased us, without your requesting me to agree to leave behind me a longer-lasting image of the image (*eidōlou eidōlon*), as if it was something genuinely worth looking at? [20]

Plotinus thus articulates in very clear terms the rationale of Plato's objections to portraiture *qua* mimesis.[21] Porphyry's account goes on to report how, in spite of Plotinus's obstinacy, his disciple Amelius succeeds in obtaining a likeness of the master, but only by virtue of a trick. He is forced to train what Mannerist theorists of art would later call a *ritrattista alla macchia*, namely, a portraitist working from memory, *a forza di memoria*,[22] in order to reach his goal: Amelius brings the painter Cartesius along to the meetings of the school, and accustoms him

> by progressive study to derive increasingly striking mental pictures from what he saw. Then Carterius drew a likeness of the impression which remained in his memory. Amelius helped him to improve his sketch to a closer resemblance, and so the talent of Carterius gave us an excellent portrait of Plotinus without his knowledge.[23]

The ambivalent attitude toward portraiture such a story exemplifies prevailed well into the following centuries and well into the Christian era, at least according to Vasari's influential account; namely, until Giotto rediscovered "the portrayal of the likenesses of living persons (*il ritrar di naturale le persone vive*), which had not been practised for many centuries."[24] As E. H. Gombrich has eloquently argued, the likelihood of this hypothesis is confirmed by an anecdote Petrarch relates in his *Senilia*. Petrarch displays under the circumstances a remarkably more tolerant attitude toward portraiture than Plotinus: he does not reject the very idea of the portrait and is not at all uncomfortable in his role as a sitter, he is only dissatisfied with the likeness the painter has been able to produce. An affluent admirer, the *condottiere* Pandolfo Malatesta, already owned a likeness of the poet. However, after meeting in person the man he had up to then only cherished in effigy, as it were, the disappointed Pandolfo deemed necessary to order another portrait. He commissioned it to Simone Martini, "truly a great artist as they go nowadays," as Petrarch praises him not without noticeable irony. "When he came to me," thus Petrarch describes the awkward situation,

> disguising his purpose, he took the liberty of sitting down with me as I was reading, for he was an intimate friend; and as he stealthily did something or other with a pen, I recognized the friendly deceit and unwillingly allowed him to paint me openly.

After witnessing Plotinus's resistance, one can better appreciate Petrarch's relaxed attitude. However, in spite of his best efforts and undeniable skills, Simone does not succeed in his endeavor, or at least so it seems to Petrarch, who converts the tale in a parable of human shortcoming: "often things attempted more aggressively succeed through carelessness, and too much eagerness kills the result."[25]

Such a commentary, however, psychologizes the reasons of the failure, which were metaphysical in Plotinus's indictment of portraiture. According to the uncompromising standards Platonism set once and for all future iconoclasms, a portrait is doomed to fail independently from the individual excellence of an artist, simply because of its ontological status, of its distance from an origin that is already conceived of as its own double. As St. Basil states the case for iconoclasm, "the image of the king is also called the king, and there are not two kings in consequence."[26] An echo of this anathema against their art is still audible in the term "volto in maestà" Italian painters used for a portrait in full face.[27] But in Petrarch the metaphysical stricture of neo-Platonism has been replaced by the painter's subjective failure to produce a convincing likeness of the sitter. In order to reproduce a face, a portrait-painter had already not just to imprint (*inprentare*) it,[28] for instance following the casting technique Cennini was teaching about the same time in his manual.[29] He had rather to capture what Petrarch himself calls—in a letter addressed to Boccaccio—the "air" of a face (*Fam.* XXIII.19).

Petrarch deserves the credit for having first introduced the category in the aesthetic vocabulary of the Renaissance. Yet, as he himself acknowledges, he was borrowing the word from the technical idiom of vernacular painters at the time: "a certain shadow and, as our painters call it, 'aria' " (*umbra quaedam et quem pictores nostri aerem vocant*). Italian painters and theorists of art drew very early on a connection between two problems they perceived to be analogous: the representation of the air of a human face, and the representation of the likewise invisible atmospheric air,—in both cases facing the problem of representing what Leopardi would later call *il bello aereo*.[30] It may be sufficient here to mention Leonardo's speculations on aerial perspective, and his unparalleled achievements in rendering the atmospheric medium through his technique of the *sfumato*.[31] But how did the name of an element come to denote the quality that marks the uniqueness of a human face?

Petrarch introduces the term "aer" in the midst of a famous discussion of imitation in writing, where he argues that the *air de famille*, which makes a son look like his father, and which should be a writer's goal to achieve, is different from the resemblance between a portrait and its model:[32]

> He who imitates must have a care that what he writes be similar, not identical [with his model] and that the similarity should not be of the kind that obtains between a portrait and a sitter, where the artist earns the more praise the greater the likeness, but rather of the kind that obtains between a son and his father. Here, though there may often be a great difference between their individual features, a certain shadow and, as our painters call it, air perceptible above all in the face and eyes (*in vultu inque oculis*) produces that similarity that reminds us of the father as soon as we see the son, even though if the matter were put to measurement all parts would be found to be different; some hidden quality there has this power (*est ibi nescio occultum quod hanc habeat vim*).[33]

Petrarch's statement that "the artist earns the more praise the greater the likeness" must be, of course, qualified in light of the letter I previously quoted and of his own authority on the way the word "aria" was used at his time. His rather narrow-minded description of the task of a portrait-painter is *prima facie* contradicted by his contemporaries' awareness of its complexity. The success of a portraitist was then already measured not so much by a standard of geometrical similarity,[34] but rather by his ability in rendering what a *connoisseur* of the time would have called the "air" of a human face. An ancient authority such as Plutarch already knew that "a portrait which reveals character and disposition (*to ēthos kai ton tropon*) is far more beautiful than one which merely copies form and feature;"[35] even more to the point, in the methodological introduction to the *Life of Alexander*, he had compared his biographical technique to that of painters who "produce likenesses from the face and from the features around the eyes (*apo tou prosōpou kai tōn peri tēn opsin eidōn*), in which the character is revealed, but pay less attention to the other details."[36] Plutarch's periphrases sufficiently identify the *nescio quid occultum*, which is perceptible above all in the eyes and their vicinity, as the proper subject of portraiture;[37] what he still misses is a formula to designate this *je-ne-sais-quoi*. George Tullie's 1684 version of the treatise on *How to know a Flatterer from a Friend* shrewdly supplements it, but only at the price of suggesting a synonymy that must not be taken for granted, as we will see: "unskilful Painters" must "content themselves with the faint Resemblance in a Wrinkle, a Wart, or a Scar," since "they can't hit the Features and Air of a Face."[38]

According to the currently accepted etymology, the Italian word *aria*, when used in reference to the human face, derives from the medieval Latin *area* or *ager*, from which the French expression *de bon aire*, meaning "of good breed," and used in reference to the successfully tamed falcon.[39] Leo Spitzer legimately questioned the validity of such an explanation,[40] although his own derivation is less than fully convincing and needs to be supplemented.[41]

There is indeed a possibility of telescoping[42] the two words "aer" and "area" in a way that neither Spitzer nor his adversaries have pursued; namely, by looking at the semantic area of ancient meteorology. The scope of this science was not limited to that of our weather forecast, but encompassed all phenomena that occur in the sky, within or without the atmosphere. The first book of Seneca's *Naturales Questiones*, for instance, is devoted to a discussion of what he generically calls "lights in the sky." Among them is the halo, namely, the bright light (*fulgor*) that surrounds stars under particular conditions of visibility. "Halo" is the Latin transliteration of the Greek word ἅλων; but Seneca provides also the translation "area," which he prefers, because more literal, to the alternative "corona," i.e., "crown," adopted after him by Pliny (*Nat.hist.* II.xxviii.): "The Greeks called such shining lights threshing-floors (*areas*) because generally the places set aside for threshing grain were round. "[43] A passage such as this may provide the missing link to connect the topical and the elemental meaning of "aria," and reconcile the two seemingly mutually exclusive derivations, from "area," on the one hand, and "aer," on the other. "Aria" would then refer to the "halo" crowning the human face—each human face, namely, and not just a saint's features. Such a suggestion will sound even more persuasive if we consider that the invention of the technical device we now call *chiaroscuro*[44] was also probably stimulated by the explanation of such celestial phenomena as the halo,[45] which Aristotle interpreted as the result of a condensation of air and vapor thick enough to reflect the light emanating from a veiled star (*Meteorologica* 371b22–26, 372b12–34).

Agnolo Firenzuola testifies that, by the mid-sixteenth century, *aria* was used "by antonomasia (*per figura di antonomasia*)" to imply "good air," being "a good sign, manifesting a clear, healthy soul and conscience," whereas *mal'aria*, or an absence of *aria*, would imply the contrary.[46] Half a century later, however, the air had already withdrawn from sight and was now hiding (or latent) in the eyes: *praecipue in oculis latet*, as Della Porta writes in his treatise *On Celestial Physiognomy*. The first book of Della Porta's work is devoted to a discussion of what he calls "character, or dignity of the aspect" (*indoles, siue character, vel aspectus dignitas*), an expression he uses to render precisely the vernacular "aria"—what "*vulgus (ariam) vocat.*"[47] However, he does not deem necessary to explain why the eyes should be the site of the air in the face, nor does so in the third book of his earlier treatise *On Human Physiognomy*, even though that book is entirely devoted to a discussion of the physiognomy of the eyes. There Della Porta reaffirms their traditional preeminence among the facial features as the site of the soul (the word "air" is not mentioned), by insisting on the topical analogy of the face as image of the soul, and the eyes as image of the face. One of the authorities he resorts to, however, provides him with a more articulated rationale. In his *Convivio* Dante had explained at length how the soul works its way through, as it were, the human face, and shapes it by virtue of its particularly subtle action. As a result,

> no face is the counterpart of another [*nullo viso ad altro è simile*]; since the final potentiality latent in the subject matter, which is in all cases somewhat different, is here reduced to actuality.

The soul, Dante continues, operates chiefly in the eyes and the mouth, because, according to the Scholastic teaching, "in these two places all three natures of the soul, as it were, have jurisdiction," namely, the vegetative, the sensitive, and the rational. Nonetheless, the eyes deserve their higher position, for

> the soul reveals herself in the eyes so manifestly that any one who gazes intently on her may know her feeling at the moment,

whereas the way she shows herself in the mouth is more opaque, "as it were, like colour behind glass (*quasi come colore dopo vetro*)."[48] Both eyes and mouth, however, deserve to be called "balconies of the lady, namely, the soul, who dwells in the edifice of the body." Dante's "graceful simile" will be echoed throughout the physiognomical literature of the Renaissance: for instance, the final recommendation in Antonio Pellegrini's dialogue *Della Fisonomia Naturale* is clearly reminiscent of Dante's passionate paean to the eyes:

> Above all, keep looking steadfastly and intently in the eyes: for it is there that almost all our affections manifest themselves, and our soul shines through them like through open windows.[49]

Later physiognomists, starting with Simone Porzio's treatise *On the colour of the eyes*,[50] and even if they were still working within the framework of the Aristotelian terminology, will feel the need to provide more than an analogical justification of the physiognomical significance of the eye. Thus the focus of attention will slowly shift onto the material texture of the whole face; as it occurs, for instance, in the idiosyncratic work of the Bergamasque physician

Giovan Battista Persona, *Noctes solitariae*, a scientific commentary in dialogues on the *Odyssey*. In the midst of a discussion on the physiology of tears, Persona goes on to argue that the countenance (*vultus*) and the eyes in particular are more "representative" of the movements of the heart than the other parts of the body, because the skin of the face is all interwoven with larger veins and arteries.[51] The trend toward a more physiologically oriented physiognomy will continue into the seventeenth and eighteenth centuries, especially due to the developments in the scientific study of muscular motion that are reflected in the physiognomical works of John Bulwer[52] and James Parsons,[53] among others. The skin, the envelope of the body ancient physiognomy had almost overlooked in favour of more permanent signs of the human character, will then be seen more and more as the canvas on which the passions of the soul spread their colors,—and not just for the blind, as Diderot's paradox would make us believe (*il y a aussi une peinture pour les aveugles; celle à qui leur propre peau servirait de toile*). The skin of the face, Jacques Pernetti writes in his *Lettres Philosophiques sur les Physionomies* (1746),

> is of a particular constitution, which one does not find elsewhere. All over the body the skin is separated from the flesh: on the face, they are so tightly bound together that one cannot separate them without tearing them apart. This makes the skin of the face somehow (*en quelque façon*) transparent, and more suited to receive and to paint for us on the outside the different colours that the various occurring movements excite.[54]

In Della Porta's treatise, the canvas is still the air, which falls, like a curtain, out of the eyes and veils the features of the face, as if it were, in his words, "another face (*altera facies*), or a transparent mask inseparable from the true face (*aut transparens persona a vera facie inseparabilis*)." The air of the face is yet another face, a face double. It is an "insignis naturae pictura,"[55] a unique copy molded out of a unique cast. And such a picture of nature cannot be reproduced by human means, it is properly inimitable by art.

The examples he brings forward show that Della Porta is here thinking this "picture of nature" on the model of the *acheiropoieta*, the images of Christ "not produced by human hands," such as, most famously, the Veronica [fig. 14].[56] More relevant to our argument, however, and historically preceding the worship of the *vera icon*, is the anecdote concernig the Mandylion, the image of Christ preserved in Edessa, which Della Porta cites in the version given by Nicephorus Callistos in his *Historia ecclesiastica*: Abgar, king of Edessa,

> sente a payntour unto Jhesu Cryste/ for to fygure thymage of oure lord/ to thende/ that at leste that he myght see hym by his ymage/

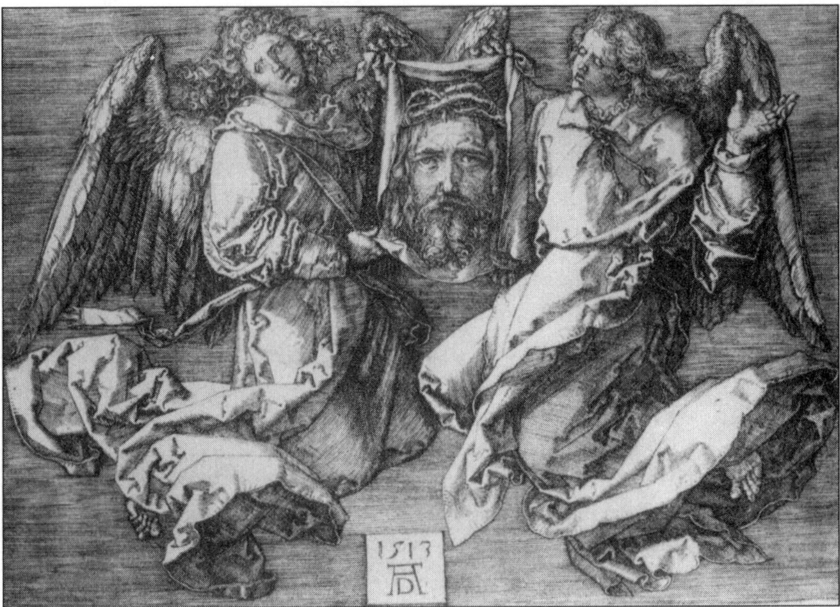

Fig. 14. Albrecht Dürer, *Das Schweißtuch der Veronika,* engraving, 1513. Reproduced from *Albrecht Dürers sämtliche Kupferstiche in Grösse der Originale* (Leipzig: Hendel 1928). McCormick Library of Special Collections, Northwestern University Library.

> whome he myght not see in his vysage/ And whan the payntoure cam by cause of the grete splendour and lyght that shone in the vysage of our lord Jhesu Cryst/ he coude not beholde it/ ne couthe not counterfeite it by no figure/ And whan oure lord sawe this thyng/ he toke fro the payntour a lynnen clothe/ and set it upon his vysage/ and enprynted the very physonomye of his vysage therin/ And sente it on to the kynge Abagar.[57]

Another version of this apologue, which even more closely anticipates Petrarch's treatment, is found by Della Porta in Plutarch's inexhaustible source, and precisely in his life of Demetrius. Unlike the previous anecdote, however, this one remains only the report of a failure:

> he had features of rare and astonishing beauty (*ideai de kai kallei prosōpou taumastos kai perittos*), so that no painter or sculptor ever achieved a likeness of him. They had at once grace and strength, dignity and beauty, and there was blended with their youthful eagerness a certain heroic look and a kingly majesty that were hard to imitate (*dysmimētos heroikē tis epiphaneia kai basilikē semnotēs*).[58]

The iconoclastic moral of the story can be phrased in Erasmus's words: "Quae sunt hominis praecipua, pictori sunt inimitabilia;"[59] a saying in which the prefix *dys-* of *dysmimētos*, suggesting a difficulty rather than an impossibility, has been replaced by the irrevocability of a condemnation. "Immane quantum illic abest hominis! Quod ex summa cute coniici potest, expressum est:" the painter can only express that which can be guessed from the outermost surface of the skin, an area well within the reach of Diderot's *aveugle*.[60]

> Those who have established Physiognomy into an Art, and laid down
> Rules of judging Mens Tempers by their Faces, have regarded the
> Features much more than the Air,

Addison remarks in his essay on physiognomy in *The Spectator*, where he defines the air as "the inward Disposition of the Mind made visible."[61] Undoubtedly, the word "air" neither figured in the vocabulary of ancient physiognomy nor grew up on the terrain of the physiognomical tradition. As we have seen, it was probably borrowed from the vocabulary of art and literary criticism, which continued in the Renaissance, as Michael Baxandall has observed, "the classical habit of metaphorical interchange" between their lexical domains.[62]

The first usage of the word "air" in the physiognomical literature I am aware of occurs in Della Porta's *On Celestial Physiognomy*, which he published almost twenty years after his better known *On Human Physiognomy*, and did not enjoy the same kind of fame as the earlier treatise. But the scope of the later work is far less ambitious and its goal chiefly polemical. It is meant as an assessment of the claims of the astrological physiognomy, whose legitimacy Porta denies in name of a physiognomy founded on the theory of temperaments, such as the one he had recently expounded in his major treatise. The work is thus a belated episode in the history of the Renaissance debate over astrology. Porta's declared polemical target is Ptolemy, whose theory that the astrological constellations influence the operation and the aspect of the sublunar world, far from remaining limited to calendars or popular works,[63] had been embraced and validated by some of the foremost thinkers of the Renaissance,[64] first of all by Marsilio Ficino in his *De vita coelitus comparanda.*

To support the case in favor of the astrological influence, Ficino evokes two very pervasive popular beliefs,—firstly, that the "vultus" of the prince has the power to affect the mood of his subjects; secondly, that the "vultus," seen or imagined by a couple at the moment of their intercourse, have the power to affect the "vultus" of the new born—and uses them as terms of comparison in an analogy, whose power of persuasion ultimately rest on the ambiguity of the Latin term:

> In a city, does not the countenance (*vultus*) of a prince, if mild and
> cheerful, cheer everybody up, but if fierce or sad, instantly terrifies

them? What then do you think the countenances of the celestials (*coelestium vultus*), the lords of all earthly things, are able to effect in comparison to these? I think that inasmuch as even people uniting to beget offspring often imprint on children to be born long afterwards not only the sort of countenances (*vultus*) they then wear but even the sort of countenances (*vultus*) they are merely imagining, in the same way the celestial countenances (*vultus coelestes*) rapidly impart to materials their characteristics. If sometimes the characteristics seem to lie hidden there a long time, eventually in their season they emerge.[65]

The legitimacy of this usage, however, cannot at all be taken for granted, as Ficino well knows. In order to keep the delicate balance of his analogies, he is forced to use the word "vultus" in a slightly innovative way. For the traditional translation of the word *prosōpa*, as used to refer to the astrological configurations, was "facies" and not "vultus."[66] He has then to redefine the two terms for his own purpose:

The countenances of the sky are the celestial figures. You may call "faces" those figures there which are more stable than the others; but "countenances" those up which change more. (*Vultus autem coeli sunt figurae coelestes. Potes vero facies illic appellare figuras caeteris ibi stabiliores. Vultus autem figuras quae magis ibi mutantur*, ibid.)[67]

In so doing, he could feel nonetheless justified, since he was coming closer, although only at the level of prosopopeia, to the linguistic propriety of classical Latin, which had drawn a clear distinction between the two terms. As Lorenzo Valla pointed out in his *Elegantiae*, in the usage of classical authors

'Facies' refers to the body: 'Vultus' rather to the soul and the will, whence it derives. For 'volo' has the supine 'vultum:' therefore we say 'with an angry and sad countenance (*vultus*)' rather than 'face (*facies*):' and on the contrary 'with a large or long face (*facies*),' not 'countenance (*vultus*).'[68]

The word "vultus" had no counterpart in Greek, according to Cicero, and was properly used only in reference to man.[69] The Greek word English necessarily translates into Latin with "face" is *prosōpon*, that is, according to the etymology accepted by most early modern linguists, "that which surrounds the eyes" (*pars quae est circa oculos*);[70] the face is thus defined solely in reference to the eyes, as a circle in reference to its center, a purely geometrical definition.[71] But none of the terms Latin uses to refer to the face reflects a similar hierarchy of the human features, neither "facies" nor "vultus" (not to mention "os," which means "mouth" in the first place, and then metonymically "face").[72]

"Facies" derives from "facere" and, as Aulus Gellius remarks (XIII.28), refers to "the entire shape and fashion, the make of the entire body" (*forma omnis et modus, et factura quaedam corporis totius*.) "Vultus" derives from "velle," as most etymologists agree; but another etymology, which makes "vultus" derive from "volvere," "to turn," was upheld as well;[73] among others, by Agostino Nifo in his commentary on the pseudo-Aristotelian *Physiognomy*. He advances this etymology as a step toward an alternative translation of "physiognomy" into Latin, a proposal that did not enjoy particular success:

> as far as words are concerned, the practitioner of physiognomy is called in Greek 'physiognomon,' but I prefer to call him with a new name 'vultispector,' and his activity 'vultispicere.' For 'vultus' is the surface that is seen, and is so called from 'volvere.' The term does not refer solely to the face, but to any visible surface.[74]

Independently from the etymology we may prefer, it is clear that the term, as opposed to "facies," identifies the pathognomical aspect of human physiognomy. "Facies" is natural and immobile, "vultus" is arbitrary and mobile.[75] In the *Triglossos*, a treatise in verses belonging to the late thirteenth century, which Petrarch repeatedly quotes in his codex of Virgil, one finds the same distinction, but with an added twist, which allows me to return to my original problem: "voltus velle notat, ast effigiem facies dat."[76]

The paronomasia *effigies-facies* clearly aims at suggesting the existence of a somehow natural continuity between the face and its effigy,[77] but only if one takes this latter term to mean something like a cast of the face: by casting, Cennini writes, "you will take the effigy or physiognomy or imprint (*la efigia, o ver la filosomia, o vero inprenta*) of every great lord."[78] On the other hand, a portrait is not simply a likeness of the human "facies," but rather an immobile "vultus." For this reason, in all likelihood, the three-quarter profile was the preferred posture of Renaissance portraits: "in the front-view the face is *turned* toward us; in the three-quarter profile it *turns* towards us," thus evoking "the illusion of movement, of action, of life."[79] But the fixity of the front-view is also an illusion, at least until the *rigor mortis* does not impose on a "vultus" the definitive features of its *facies hippocratica*. Ironically, while still breathing, one cannot show one's full face without performing a volte-face.

I would like to sum up the previous discussion by suggesting that the air is the halo created by a face, is the niche, so to speak, a face carves for itself in the surrounding air by overcoming its own inertia. This means in turn that only the expression "aria del volto" is proper, as Italian is the only Romance language[80] to have preserved up to the present time a word directly coined on "vultus" and referring to the human face.[81] Only a "volto" can put on airs. One can still hear an echo of the volitive slant of "vultus" in the first two lines of a poem by Guido Cavalcanti: "Chi è questa che ven, ch'ogn'om la mira,/ che fa tremar di chiaritate l'âre" (*Rime* IV.1–2).[82] The imperious countenance of the face makes the air tremble as it cleaves it. But the distinction between "faccia" and "volto," as well as that between both these terms and "aria," will be lost very early on. Leonardo already uses the two terms indifferently; in 1550 Annibale Raimondo titles a chapter of his treatise *On the Art of Naming* "De la faccia, over volto;"[83] and later painters will consistently use the expression "aria di testa" "in order to express the aspect of the faces (*per esprimere l'aspetto de' volti*)," as Baldinucci records in his *Vocabolario toscano dell'arte del disegno*.[84] The same author gives a definition of "fisonomia," which is still unaware of the distinction between features and air Addison will draw few years later: "Arte, per la quale dalle fattezze del corpo, e da' lineamenti e aria del volto, si conosce la natura degli uomini."[85] The Italian word "fattezze" is a true cognate of English "features," and both in turn spring from the same root, Latin "facere;" even more immediately, one might suggest, both terms derive from Gellius's usage of "factura" in the passage I quoted above: *forma omnis et modus, et factura quaedam corporis totius.*

Baldinucci's definition seems to imply that, by the time he compiled his dictionary, "air" and "face" had become practically synonymous. The Abbé Pernety testifies that a similar process had come about in the French language by the second half of the eighteenth century: "On prend souvent l'air pour le visage même; on dit alors d'une personne: elle a l'air modeste, un air triste, un air de douceur, qui enchante &c."[86] Leopardi reports similar circumstances for Italian in his *Zibaldone*: "Si dice tutto giorno *aria di viso, fisonomia* ec. e *la tal aria è bella, la tale no*, e *aria truce, dolce, rozza, gentile* ec. ec."[87] In the same entry Leopardi gives his own explanation of the shift in the meaning of "aria," which he grounds in yet another ingenious etymological derivation:

> one has given the name of air [. . .] to this general signification of a physiognomy precisely because, consisting of very subtle relationships with the immaterial qualities of man, it cannot be determined and is almost an airy thing.[88]

Leopardi's explanation directly follows from his consistent refusal of the theory of beauty as *convenienza*—or symmetry—of parts.[89] Leopardi rejects the very idea of an "absolute" beauty in name of a theory of "ordinary" beauty, as it were, which he considers dependent on, and subordinate to, "significazione." He most clearly spells out his objections in an entry on August 17, 1821:

> the signification of the physiognomies [. . .] is completely different
> from absolute beauty, and is nothing else than a relationship set by
> nature between the inside and the outside, between the habits etc.
> and the figure; this signification, I say, is a foremost part of beauty,
> is one of the capital reasons whereby this physiognomy produces in
> us a feeling of beauty, and that one the contrary. No physiognomy
> can be beautiful, which does not signify something pleasant [. . .];
> and a physiognomy that means something unpleasant is always ugly,
> even if most regular.[90]

Leopardi proceeds to admit that, "ordinarily," one can safely admit a correspondence between external and internal regularity, but this *consentement préétabli* must be qualified by the historical awareness of the decay of mankind from its natural state:

> since the inside of man loses its natural state, and its outside more
> or less preserves it, the signification of the face is mostly false; but
> even though we know it, we are nevertheless attracted (and some-
> times even moved) by a beautiful face, when we see it. And we
> believe that such an effect is completely independent from the
> signification of that face, and derives from a completely separated
> and abstract cause, which we call beauty. But we flatly delude our-
> selves, because the particular effect of human beauty on man [. . .]
> always essentially derives from the signification it contains, a
> signification that is completely independent from the sphere of beauty,
> and neither abstract nor absolute at all.[91]

His vision of habit as creating a "second nature" dictates Leopardi's diagnosis of the loss of expression civilization brings about. If it is true that the signification of a physiognomy mostly derives from habits, and that these "put a physiognomy into action, and give to it its representativeness,"[92] then the same civilization that created a code for the expression of emotions paradoxically brings about their obsolescence: first and second nature collide, with the result of "almost destroying [. . .] the principal distinction that nature has set between animate and inanimate things, between life and death, namely, the faculty of movement."[93]

In spite of its unrelated etymon,[94] "countenance" is the best credited term to translate "vultus" into English.[95] Probably the first physiognomical work ever published in English, Thomas Hill's *Contemplation of Mankinde* (1571) introduces "countenance" as the equivalent of "vultus," and adds another quite ingenious but very much unlikely etymology of "vultus" to those I have previously discussed:

> The face is often taken, and that simply, for the naturall looke of
> any: but the countinance [*sic*] signifieth the qualities of the mind
> [. . .] In a man the face remaineth, but the countenance doth alter: so
> that the countinance [*sic*] is named of the Latin word *volando*, which
> properly in English signifieth a flying or vanishing away.[96]

Two centuries later, the translators of Johann Albert Bengel's *Gnomon*, an
eighteenth-century commentary on the New Testament (in a gloss to Matth.
XVI, 3: *"prosōpon tou ouranou"*), use "countenance" and "face" in order to
draw exactly the same distinction Ficino had drawn between "vultus" and
"facies":

> *vultum coeli*) non, *faciem. Vultus* hominis variat, *facies* semper est
> eadem. Prosopopoeia [. . .] modo.
>
> *the countenance of the sky*) not *face*. A man's *countenance* varies;
> his *face* is always the same. An instance of Prosopopoeia.[97]

Now I can go back to my promise of submitting a new translation for
dysōpia that may replace "compliancy" and "bashfulness," even if the latter
term does not lack, like the former, a physiognomical dimension. It derives from the verb "to abash," which would in turn derive from a supposed Italian "baïre," "to astound," formed by way of onomatopoeia from
bah!, "a natural exclamation of astonishment."[98] The verb would hence
imitate the monosyllabic sound produced by a "quick opening and closing
of the mouth," which Kant lists among the universally understood ges-
tures [fig. 15].[99] But the Greek term unquestionably refers to a changed
cast of the eyes. This knowledge is the threshold of Plutarch's interpretation. He writes:

> as dejection (*katēpheia*) is defined as pain that makes us look down,
> so when modesty yields to suitors to the point where one does not
> even look them in the face, it is termed "compliancy."[100]

Plutarch takes the etymological sense of the word to be "to become incapable of facing someone," "to be unable to return somebody's gaze." I
prefer the alternative derivation, according to which the face loses its composure and the gaze its directness because of the other's disregard. For this
reason, I submit, "discountenance" would be, short of a transliteration, the
best approximation to *dysōpia*: somebody's else disregard put us out of
countenance.[101] In losing our countenance, however, we do not necessarily
lose our face.

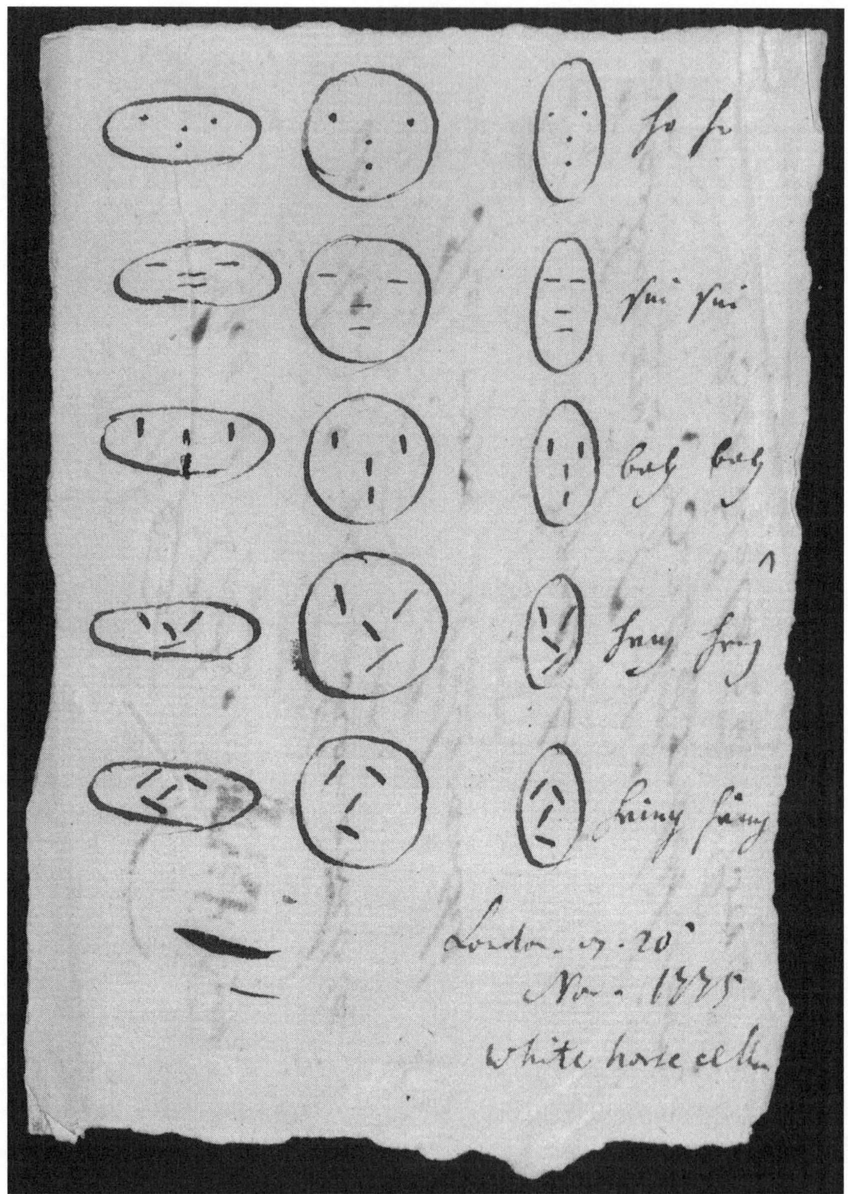

Fig. 15. Georg Christoph Lichtenberg: "ho ho/sni sni/bah bah/heng heng/heing heing" Cod. Ms. Licht. V, 10c: Bl. 1 r. Photo Courtesy Niedersächsische Staats- und Universitätsbibliothek Göttingen.

Chapter 4

Air and *Aura*

To breathe is a fulfilling of desire.

Wallace Stevens

The surface of the water mirrors the motions of the air, as the countenance the passions of the soul.[1] According to Isidorus of Seville, who often provides the best clues to understand the transition from the classic to the early modern usage of a Latin term, *vultus* is also appropriately used to refer to the changing surface of sky and sea,

> for the sea often is set in various motions by the blowing of the winds, and the countenance of the sky (*coeli vultus*), as well, changes from bright to dark and from serene to cloudy, as does the countenance (*vultus*) of men along with the change of their minds.[2]

For this reason, he adds, one calls "vultuosi" those whose countenance often varies.[3] Isidorus's analogy may help us understand how meteorological categories eventually enriched the vocabulary of physiognomy. The exchange between the two lexical domains could also be furthered by the ancients' practice of personifying the winds and most other meteors by, literally, lending them a face.

Faces with inflated cheeks are customarily depicted on the margins of Medieval wind diagrams and Renaissance maps as blowing from the different points of the compass [fig. 16].[4] In his treatise *On Painting* Alberti recommends the use of such devices in order to account for the seemingly unnatural postures of inanimate things, such as hair, leaves, clothes, when represented in a state of motion: "for this reason it will be fit to put in the painting the face of the wind Zephyr or Auster blowing among the clouds (*per questo*

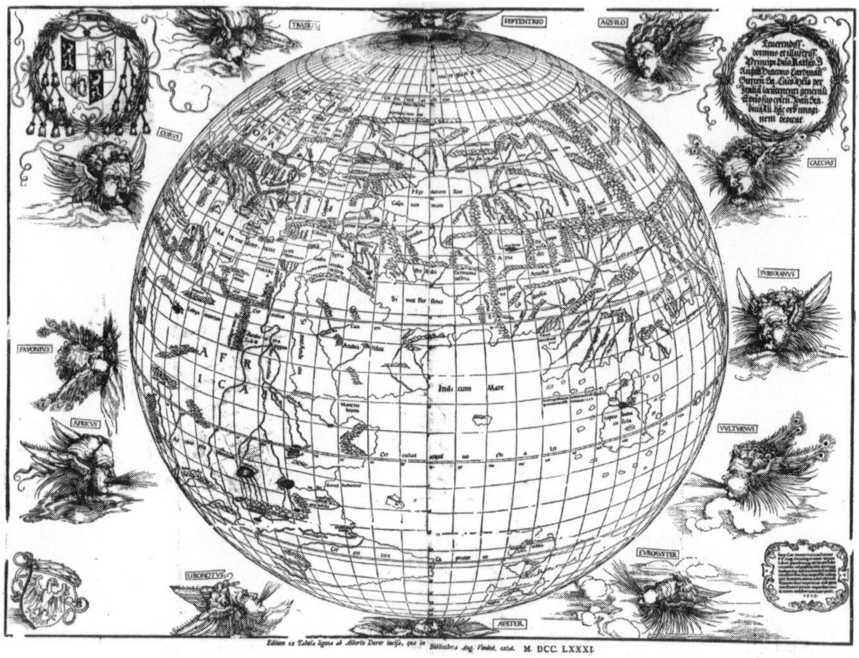

Fig. 16. Albrecht Dürer and Johannes Stabius, *World Map*, 1515. Woodcut. Photo Courtesy The Newberry Library, Chicago.

starà bene in la pittura porvi la faccia del vento zeffiro o austro che soffi tra le nuvole)."[5] This technique of representation inspired the composition of such masterpieces as Botticelli's *Spring* and *Birth of Venus*, where the winds appear as full-bodied actors on the stage [figs. 17–18]. But this ingenious shortcut will be deemed no longer necessary in High Renaissance art, when the regained confidence in the illusionistic power of painting will discard such artifices as childish.

Shortly thereafter, Francis Bacon will standardize the nomenclature of the winds in his *Historia ventorum,* not an insignificant part of his overall attempt to reform the ancient system of knowledge: for winds had meant wings to humankind, as he writes,[6] but especially to the Elizabethan ships in their race for new shores and new markets. Bacon's stated purpose is to bring order in the maddening plethora of names inherited from ancient mythology, by adopting the coordinates on the plane of the horizon as the only element for their identification [fig. 19].[7] The consequences of this renaming were far-reaching and may explain why in later poetry, even in those Romantic poets otherwise so sensitive to the calling of "airy tongues," we only find a handful of names of winds, to which capitalization can hardly lend flesh and blood.[8]

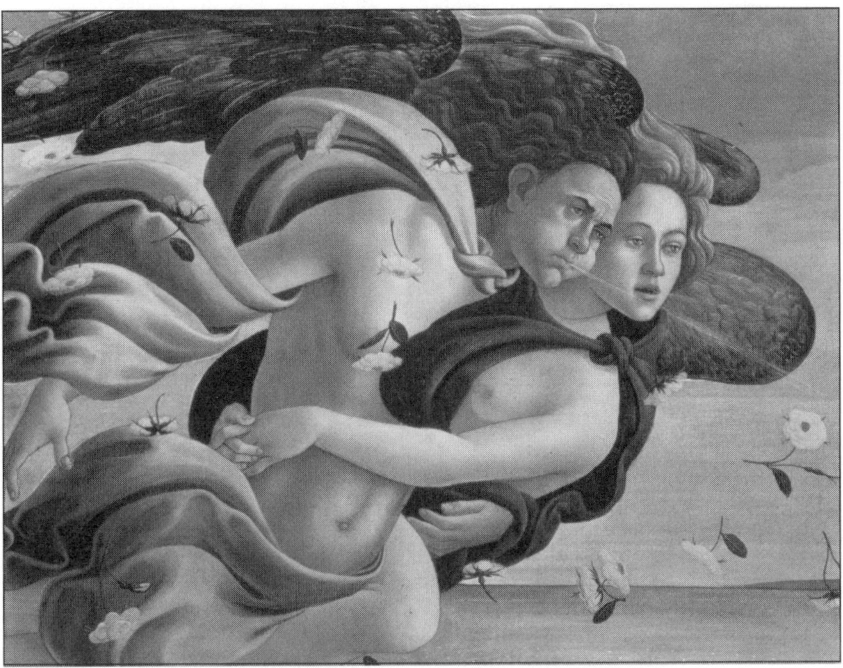

Fig. 17. Sandro Botticelli, *Nascita di Venere* (detail), Florence, Uffizi. Photo: Archivi Alinari-Giraudon, Firenze.

On the other hand, "a breath of wind (*un fiato di vento*)," as Dante still bespeaks the ancients' view, "changes name as it changes side (*muta nome perché muta lato*)."[9] A spurious treatise of the Aristotelian *corpus* (the standard treatise on winds being the work of Theophrastus) provides a complete list of the Greek names for the winds, including also regional variants.[10] As in the case of the anatomical terms, Renaissance writers emphatically lamented the loss of such a wealth of names—thus Firenzuola praises in hyperbolical terms the "boldness" of the ancient "generation," which had made "of each breath of wind a name and a difference," over the modern (*d'ogni soffiamento di vento, fa vn nome, fa vna differenza quella audace generatione*).[11]

Yet not even the ancients ever made a similar attempt to classify the breezes. Aristotle spells out in the following terms the difference between "wind" and "breeze" in his treatise *On the Cosmos*: "The breath (*pneuma*) that breathes in the air we call *wind* (*anemos*), and the breath (*ekpnoē*) that

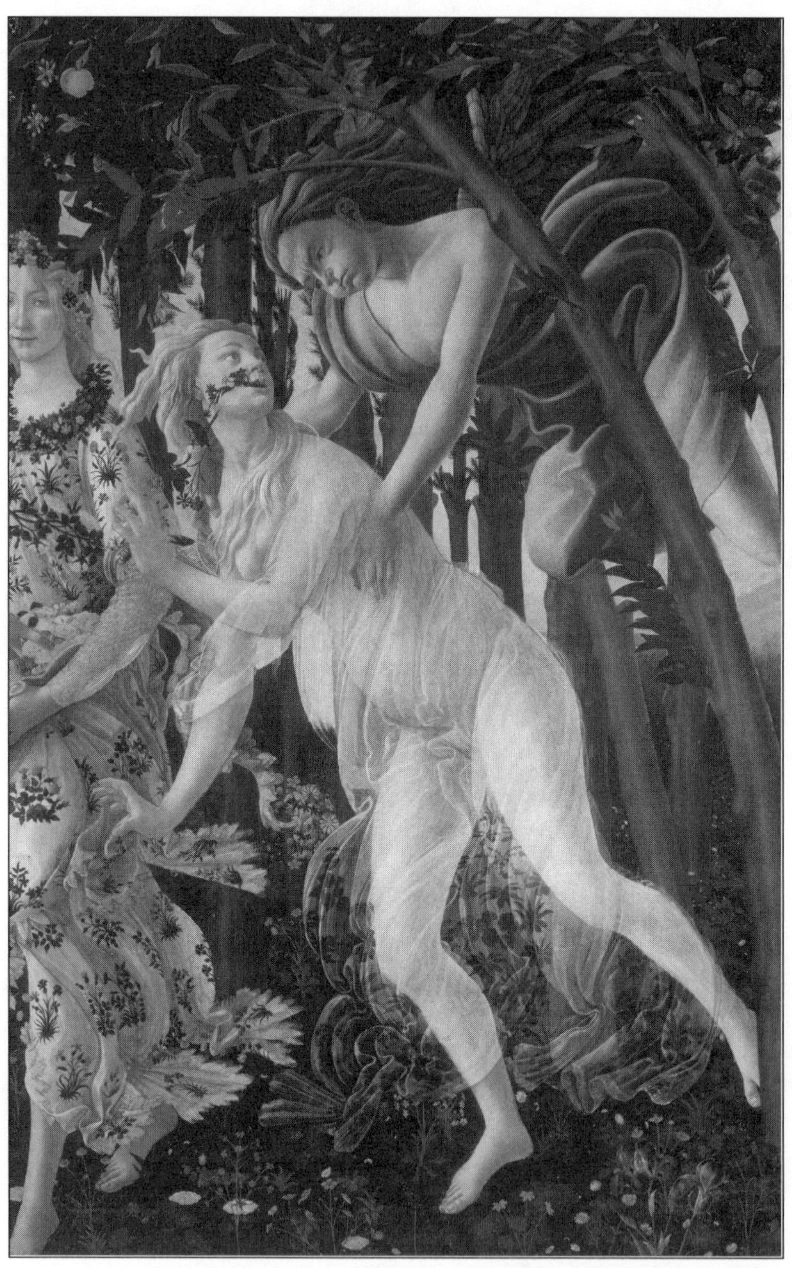

Fig. 18. Sandro Botticelli, *Primavera* (detail), Firenze, Uffizi. Photo: Archivi Alinari, Firenze.

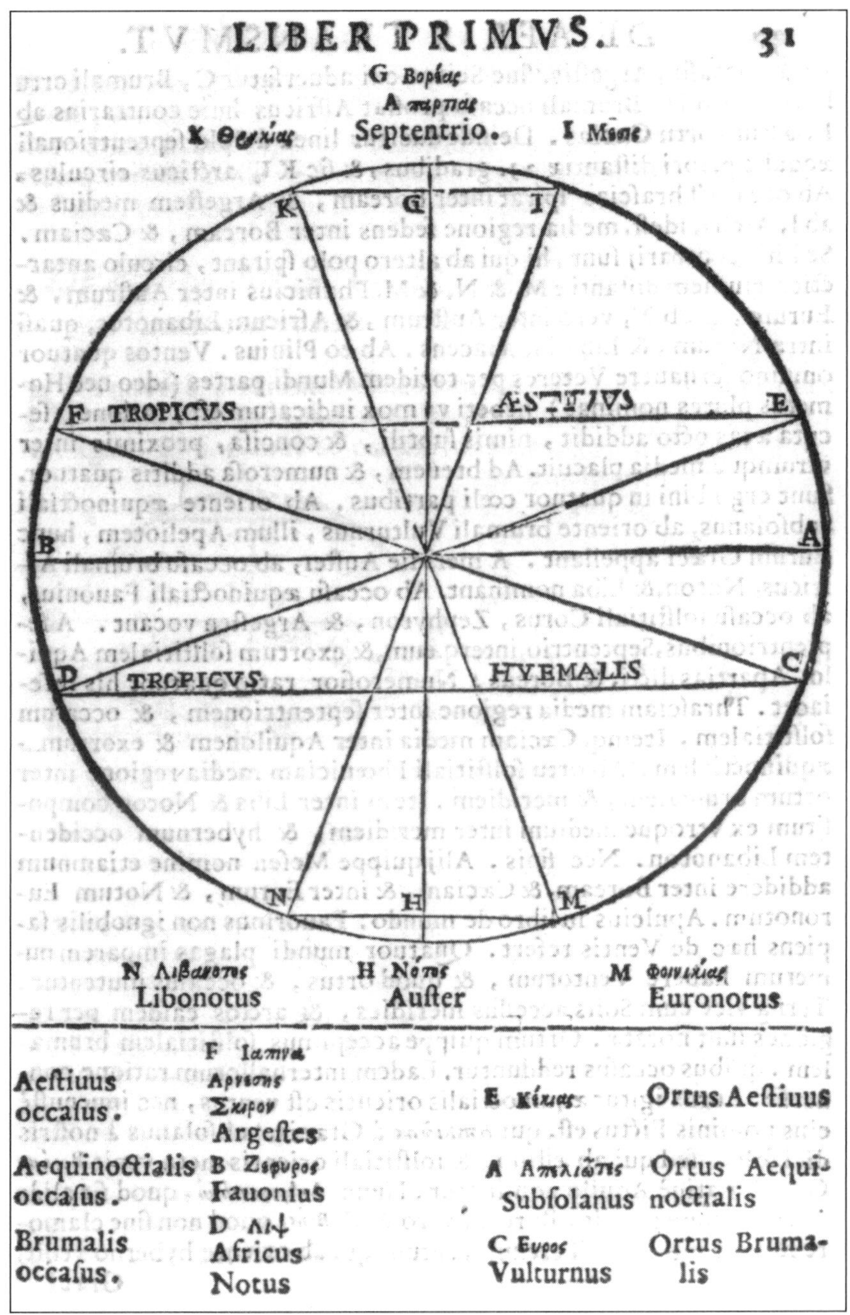

Fig. 19. Diagram of the winds, from Giambattista Della Porta, *De aeris transmutationibus,* Roma 1610. Special Collections Research Center, University of Chicago Library.

comes from moisture we call *breeze* (*aura*)."[12] Then he moves on to classify the winds according to their directions, and does not pursue further the subject of the "aura." There has probably never been the need for such a classification, given that an aura was neither a sign unequivocal enough of the forthcoming changes in weather, nor a current of air strong enough to fill the sails of a Greek or Roman sailboat.[13]

Rather than by its direction, the aura is defined by its intensity: it is a more impalpable, a subtler spirit than the wind, and a gentler interlocutor to the poet. To talk to the wind is a senseless endeavour, an *adynaton*, in the terms of classical rhetoric,[14] and a still current Italian adage: "parlare al vento," which amounts to "wasting one's breath." Petrarch uses also the less common form: "I write in the wind," (CCXII) when he describes the hopelessness of his longing for the healing "summer breeze" (*aura estiva*). On the other hand, the poet can talk to the aura without fear of having to eat his own words. The sound of a sigh is close enough to that of an aura, and this consonance sets the tone of the *Canzoniere* from the very first line of the first sonnet: the reader is invited to listen to the sound of the poet's sighs, a sound that frequently takes the shape of the name "Laura:" "I move my sighs to call you near" (*io movo i sospiri a chiamar voi*), so opens the fifth sonnet, in which the name of the beloved is for the first time anagrammatized.

> By "air" the ancients understood

> that thin, fluid, diaphanous, compresible and dilatable Body in which we breath, and wherein we move, which envelops the Earth on all sides to a great height above the highest Mountains; but yet is so different from the *Æther* [or *Vacuum*] in the intermundane or interplanetary Spaces, that it refracts the Rays of the Moon and other remoter Luminaries.[15]

Boyle's definition, which I excerpt from his *General History of the Air*, makes clear that the air in the ancients' cosmos is confined to the sublunar world, to the atmosphere proper. The Greek word ἀήρ, which has passed to English through its transliteration into Latin,[16] always signifies in Homer, as in the Presocratics philosophers, mist.[17] At an early stage of its linguistic development, the Greek air is, by antonomasia, a misty air.[18] In this usage we see already reflected the ambivalence in our attitude toward the air that Hamlet later so eloquently expressed:

> This most excellent canopy the air, look you, this brave o'erhanging, this majestical roof fretted with golden fire—why, it appears no other thing to me than a foul and pestilent congregation of vapours.[19]

The air we breathe makes up the sphere of life in the ancients' view of the cosmos, what the Greeks called ὁ περιέχων ἀήρ—"the circumambient air" of a seventeenth-century English poet, Sir John Suckling,[20] being a faithful translation of the Greek expression.[21] In his 1869 essay *The Queen of the Air*, John Ruskin takes Athena to be the representative of "the ambient air"[22] in Greek mythology and thus describes "the several agencies of this great goddess:"

I. She is the air giving life and health to all animals;

II. She is the air giving vegetative power to the earth;

III. She is the air giving motion to the sea, and rendering navigation possible;

IV. She is the air nourishing artificial light, torch or lamplight; as opposed to that of the sun on one hand, and of *consuming* fire on the other;

V. She is the air conveying vibration of sound.[23]

Independently from the accuracy of Ruskin's account from the point of view of "scientific mythology,"[24] his distinction between a "life-enhancing" (under which category I propose to collect Nos. 1, 2, 4) and a "communication-enhancing" (Nos. 3 and 5) agency of the air is still useful and may help us better analyze the ancients' view. I start by examining the former, or "life-enhancing" agency.

The standard poetic epithet for the air is "vital," up to Thomas Traherne's epigrammatic formulation of the economics of breath in his poem "The Circulation:" *No Man breaths out more vital Air,/ Than he before suckt in,*[25] and to Gerard Manley Hopkins's comparison of the "nursing element" to the "Blessed Virgin."[26] In the scientific terms familiar to the ancients, taking again as a point of departure the Aristotelian *corpus*, and in particular a passage from the *De Generatione animalium* (767a31–34), one can see that Aristotle assumes the existence of a correspondence between "the bodily condition of a person" (*hē diatesis tou sōmatos*) and "the blend (*krasis*) of the surrounding air," the latter providing the first nourishment of a body.[27] Pursuing this line of thought even further, the medical writer known as Anonymus Londinensis imaginatively likens men to plants, "for as they are rooted in the earth, so we too are rooted in the air by our nostrils and by our whole body."[28] The Hippocratic treatises *Airs, Waters, Places* and *Breaths* most forcefully draw the attention of the medical practitioner on the quality of the air, which is bent to affect the bodily constitution of human beings as they are constantly exposed to its influence. Yet the air has an even closer relationship to the soul than to the body, as the soul embodies the principle of animal movement within the body. A laconic Latin epigraph beautifully states the relationship of the soul to its element: "The earth has the body, the stone the name, and the air the soul" (*Terra tenet corpus, nomen lapis atque animam aer*; *Carm. epig.* 1207, 1). According to the doxographical tradition, the pre-Socratic philosopher Anaximenes considered air the origin of the universe, which it had produced by way of rarefaction and condensation; but the first philosopher who explicitly identified the soul with the air was Dionysius of Apollonia; he was followed by the Stoics, who then made current the distinction between *aēr* and *pneuma*, namely, external and internal air.[29]

As we now look at its "communication-enhancing" agency, the function of the air as the medium of communication was established beyond doubt, as long as the belief that "there is nothing that is empty of air," which the author of the treatise on *Breaths* firmly states,[30] remained unchallenged. But the final refutation of the *horror vacui* irreparably damaged the reputation of the magician and the astrologer, as the action at a distance, yet through a medium, was the assumption on which both the power of the magician to influence the course of nature and the power of the stars to influence the sublunar world were dependent. John Evelyn, while denouncing the pollution of the air of London in his essay *Fumifugium: or, the Inconvenience of the Aer, and Smoake of London Dissipated*, a text first published in 1661,[31] concedes that one of his reasons for concern was precisely the unpredictability of "the celestial influences" under the new atmospheric conditions, as they are bent to be

> so much retarded or assisted, and improved through this omnipresent, and, as it were, universal Medium: For, though the Aer in its simple substance cannot be vitiated; yet, in its prime qualities, it suffers these infinite mutations, both from superior and inferior causes, so as its accidental effects become almost innumerable.[32]

Yet neither discontinuity nor mutation could intervene to affect the agency of the air as "conveying vibration of sound," as the medium of the human voice,—we come thus to the last item in Ruskin's list. Chaucer wonderfully adapts the physical theory of the diffusion of sound to his poetical vision of the spreading of Fame:

> Soun is noght but air y-broken,
> And every speche that is spoken,
> Loud or privee, foul or fair,
> In his substance is but air.[33]

However, while seemingly neutral, the air, by being "the mansioun/Of every speche, of every soun,/Be it either foul or fair,"[34] is necessarily liable to produce opposite effects, as Fame can always turn into ignominious Infamy. Hence the three witches of *Macbeth* are properly redressing Chaucer's alternative when they echo it in their enigmatic tautology: *Fair is foul, and foul is fair*. Once again, a Shakespearean character denounces the ambivalence of the air, which the ambiguity of the human word merely mirrors. When Anthony Hecht warns us against the double agency of language and its vehicle: "This atmosphere, which is our medicine,/By its own delicacy kills,"[35] we cannot fail to hear in his lines another echo of Macbeth's tragedy. The castle of Macbeth has an air, according to King Duncan, which "nimbly and sweetly recommends itself unto our gentle senses," and Banquo confirms the remark by observing that "where the temple-haunting martlets most breed and haunt [. . .] the air is delicate."[36] We know that this delicate air will soon turn into a deadly one. As Hamlet cannot but remind Horatio with his "dying voice," we are forever condemned to draw our breath "in pain."[37]

The air, as we have seen, is the universal medium of communication in the pre-Torricellian and pre-Boylean world, namely, before the discovery of the atmospheric void would introduce a solution of continuity in the *plenum* of nature; the aura is a modification of this medium.[38] Tommaso Campanella, still writing before the caesura, can argue in his treatise *On the Sense of Things and on Magic* that "the air is like a common soul, which helps all and through which all communicate" (*l'aria sta come anima commune che a tutti aiuta e per cui tutti comunicano*).[39] Through the air, by virtue of its elasticity and transparence, not just sounds, but also light, odors, cold, and heat, are instantaneously transmitted at a distance that is dependent only from the variable permeability of the medium under different atmospheric conditions.[40] But the air is above all the medium of the human voice. Whenever we speak, we also produce an aura, a vibration of the air that reaches the "aures," or ears, of the listener,—another punning possibility offered by the name "Laura:" "the ear," Albertus Magnus writes, "hears nothing but that which communicates with the trembling air" (*auris non audit nisi quae communicant cum aere tremente*).[41] Dante heightens, as it were, the natural treble of a sigh, when he talks of the unbaptized's sighs in limbo, "which caused the everlasting air to tremble" (*che l'aura etterna facevan tremare*; *Inf.* IV.27). In his commentary on Dante's *Inferno*, Boccaccio criticizes the usage of "aura" in the line I just quoted, by pointing out that aura is "a gentle movement of the air" (*un soave movimento d'aere*, Isidorus has "lenis motus aeris," *Etym.* XIII.xi.17). Such a movement would be, of course, out of place in the Inferno, where every movement is "impetuous and annoying" (*impetuoso e noioso*). Therefore the generic "aere atterno" would better describe the underground atmosphere the damned are condemned to breathe.[42] On the other hand, no question of propriety can be raised over the pairing "aura dolce" in *Purg.* XXVIII.7, also in consideration of its classical ancestry.

"Dulcis aura" is indeed a Virgilian locution.[43] Therefore it is not surprising that, at the moment when he has to enter Eden without the help of his guidance, Dante feels the need to confirm what he had said to Virgil at the moment of their first meeting: that he was the only model of "the beautiful style" for which Dante had been honored (*tu se' solo colui da cu' io tolsi/lo bello stilo che m'ha fatto onore*). It is thus both a sign of gratitude, and a gesture of self-reassurance if he moves his first steps still in the path of his master. Dante does so by quoting him, a high example of poetic *pietas* at the moment when Virgil is forced back into the silence whence he came (remember the way he is introduced in canto I of *Inferno*: "one who seemed faint because of the long silence," *chi per lungo silenzio parea fioco*): "un'aura dolce" is the soft breeze that still moves the branches of the forest in the lost *father*land (here a necessary choice) of humankind, the "breathing garden" of Eden.[44] Dante is granted access to the earthly paradise as a springboard toward the celestial, and there he finds, unchanged in direction and intensity, the same breeze that was blowing in Eden on the inaugural day of human history, when our progenitors heard "the voice of Lord God walking in the garden in the cool of the day." This is the reading of the Authorized Version; but Jerome's choice to render the Hebrew *ruach* is here precisely *aura*: "vocem Domini Dei deambulantis in paradiso ad auram post meridiem."

Yet Dante was certainly aware of another interlinguistic echo in using the word "aura" at this particular juncture of the *Divine Comedy*. The Provençal usage of the word "aura" had to be very much in his mind, especially following his recent meeting with Arnaldo Daniello in the Round of the Lustful in canto XXVI. There he had let Arnaut speak Provençal, certainly a sign of high deference for the poet who had been, according to Dante, the "best artificer" (*miglior fabbro*) of his own mother-tongue. Dante's esteem is confirmed by the *De vulgari eloquentia*, in which he calls Arnaut the finest poet of love, a judgment shared by Petrarch, as well. Petrarch calls Arnaut the foremost among a group of troubadours participating in the *Triumph of Love*, "Arnaut Daniel,/the great master of love, who still honours his country/ with his strange and beautiful style (*il primo Arnaldo Daniello,/gran maestro d'amor, ch'a la sua terra/ancor fa onor col suo dir strano e bello*; IV. 40–42.)"

The line by which Arnaut introduces himself in canto XXVI, "I am Arnaut, who, going, weep and sing" (*Ieu soi Arnaut, que plor e vau cantan*), is an allusion to a line from the famous *tornada* that opens: "Eu son Arnauz c'amas l'aura."[45] Arnaut's association of his own name to the word "aura" is particularly significant, since it occurs as the first *adynaton* in a series that has become almost his signature: "I am Arnaut, who piles up the breeze/And chases the rabbit with the ox/And swims against the swelling tide," activities as senseless as that of "writing in the wind."[46] Petrarch is probably mindful of both Arnaut and Dante when he combines the "crying and singing" of Dante with the unlikely "ox"-hunting of Arnaut in his poem CCXXXIX, 35–36: "weeping and singing our verses,/We shall go with a lame ox hunting the aura (*lagrimando e cantando i nostri versi/e col bue zoppo andrem cacciando l'aura*)."[47]

Arnaut's predilection of the word "aura," so much so that he has been credited for suggesting to Petrarch his *senhal*,[48] is confirmed by the incipit of another song, "L'aura amara," which Dante singles out in the *De vulgari eloquentia* as an outstanding example of love poem. Dante must have had in mind the opening stanza of this poem in his description of the selva of Eden, for he proceeds *e contrario* to describe the soothing effects of the "aura soave" that blows in Eden, denying one by one those caused by the "aura amara" and so vividly portrayed by Arnaut. Let us first read Arnaut:

L'aura amara	The bitter breeze
fa.ls broils brancuz	lightens the leafy branches
clarzir	that the sweet one
que.l dous' espeis' a foils,	thickens with leaves,
e.ls les	and the happy
becs	beaks
dels aucels ramens	of birds
te balbs e muz	it holds stammering and mute.[49]

Now Dante:

le fronde, tremolando, prone tutte
quante piegavano a la parte
u' la prim' ombra gitta il santo monte;
non però dal loro esser dritto sparte
tanto, che li augelletti per le cime
lasciasser d' operar ogne lor arte;
ma con piena letizia l' ore prime,
cantando, ricevieno intra le foglie,
che tenevan bordone a le sue rime.

the trembling boughs—they all
bent eagerly—inclined in the direction
of morning shadows from the holy mountain;
but they were not deflected with such force
as to disturb the little birds upon
the branches in the practice of their arts;
for through the leaves, with song, birds welcomed
the first auras of the morning joyously,
and leaves supplied the bourdon to their rhymes.[50]

On the one hand, a still life, or a *natura morta*, is the work of the "bitter aura;" on the other, an *aria* in the musical sense of the word, sung by the birds and accompanied by the leaves, which play like an aeolian harp to the tune of the "sweet aura."[51]

The words of another troubadour, Cercamon, may help us temper the dissonance of these two imageries: he refers to the change of breeze announcing the beginning of winter as the moment "when the sweet aura turns bitter/ and the leaf falls from its bough/and the birds change their language" (*quant l'aura doussa s'amarzis/e.l fuelha chai de sul verjan/e l'auzelh chanjan lor latis*).[52] In the Aristotelian universe, as we have seen, the aura is a modification of the air, and air, like all the other elements, is a neutral substance in itself, capable of assuming contrary qualities at different moments in time. But the seasonal change does not occur in Dante's Eden, which its Creator has located on the top of the high mountain of Purgatorio precisely to save it from this fastidious alternance of "sweet" and "bitter." For the aura that blows in Eden is not the same that blows on the earth. Dante is using the word equivocally, as Matelda's explanation will make clear: the aura in Eden is the result of the circular, unwavering motion of the atmosphere, which turns around the earth in synchrony with the superior spheres, whereas the terrestrial is a product of the humid exhalations of the earth.

Yet the ambivalence of the mundane aura, as opposed to its paradisiacal counterpart, is preserved by the verb Dante chooses to describe its action. The breeze *wounds* his forehead as he moves into the forest: "A sweet breeze, which did not seem to vary/within itself, was wounding my brow/but with no greater force than a kind wind's" (*Un'aura dolce, sanza mutamento/ avere in sé, mi feria per la fronte/non di più colpo che soave vento*).[53] The word might be discounted as an hyperbole, and its strength is undoubtedly attenuated by the following comparison,—although by now we know that even a "soave vento" is stronger than any aura, and this one is strong enough to provide a *basso continuo* to the song of the birds. Yet the verb is more intuitively used for the action of the light on the eyes in *Inferno* X.69, where Guido Cavalcanti's father periphrastically refers to life as that condition in which "the sweet light" (*lo dolce lume*) would still *wound* the eyes of his son. Hence probably Dante's trope.[54] But Petrarch himself must have wondered at this daring extension of the proper meaning of the word, and has given us this time a retroactive clue, a posthumous rationale to Dante's choice. He writes in the sonnet CXCVI:

> *L'aura serena che fra verdi fronde*
> *mormorando a ferir nel volto viemme,*
> *fammi risovenir quand'Amor diemme*
> *le prime piaghe, sì dolci profonde.*

> The calm breeze that comes murmuring
> through the green leaves to wound my brow
> makes me remember when Love gave me
> the first deep sweet wounds.[55]

The action of the aura is assimilable to the action of love by virtue of the ambivalence of their effects: like Achilles's spear, they are the carriers of both wounding and healing. We may conjecture that the same assimilation was at the core of a treatise on love, now lost, by Giovan Giacomo Calandra, of which we know only what Mario Equicola reports in his treatise *Di Natura d'Amore*: that it was titled *Aura*,

> with an allusion to the origin of the Greek name, for what we call 'love' means 'breathing,' and 'sighing' (*un libro nominato Aura alludendo alla origine del nome greco, che afflare, e spirare dinota, quel che noi amore diciamo*).[56]

The first occurrence of the word *aura* in the Septuaginta is in 1 Kings 19,12, where it refers to the form under which God manifests Himself to Elijah on mount Horeb. Wind, earthquake, and fire anticipate in turn, rather than announce, the appearance of God; for, as the anaphorical structure of the text emphasizes, in none of them is God present:

> the Lord passed by, and a great and strong wind rent the mountains, and brake in pieces the rocks before the Lord; but the Lord was not in the wind: and after the wind an earthquake; but the Lord was not in the earthquake; and after the earthquake a fire, but the Lord was not in the fire.

When it finally takes place, the theophany thus comes rather as an anticlimax: Elijah answers to the calling of a *phonē auras leptēs*, which is rendered in the Vulgata as "sibilus aurae tenuis," and in the Authorized Version as "a still small voice." The Hebrew word is, in this case, דממה, *demamah*, which refers to the calm upon the wind following the tumultuous manifestations of elemental powers.[57] This atypical theoph*o*ny, which is described in the same terms in Job 4,16 (*et vocem quasi aurae lenis audivi, auran kai phonēn ēkouon*),[58] lays bare the power of God at its most understated expression, and is probably meant as an alternative to an overly sensuous and hyperbolical understanding of the divine action.[59] Pseudo-Dionysius mentions the "aura" among the names of God in his treatise *De divinis nominibus*,[60] where he summons it out of the biblical text just before enlisting the epithet that sounds as the very disproval of his endeavor: the Anonymous, *to anōnymon*. As God cannot be called by any truly proper name, Aura can nevertheless claim for itself a high rank in the hierarchy of His names, for it names the sensuous manifestation of His power that is the least perceivable: the breathing of the breeze.[61]

If in this development of the Judeo-Christian tradition the name "aura" is thus acknowledged the status of a divine name, in the Greek-Roman tradition its emergence as such is already marked by the stigma of decay. In Ovid we witness at the same time the apotheosis of the aura and its grammaticalisation.[62] The two versions of the myth of Cephalus and Procris that he recounts in the *Ars amatoria* and in the *Metamorphoses* agree in making of equivocation the cause of Procris's unwarranted death, and "aura" is the *vox ambigua*[63] that induces her undoing.

In both versions Cephalus unwillingly provokes the jealousy of Procris by invoking the "wandering aura (*mobilis aura*)"[64] to come and relieve him of the summer heat. Procris misinterprets at first her husband's invocations to the aura as a declaration of love to a nymph by the same name,[65] and fears what is, in truth, the rivalry of a mere "nothing," "a name without a body."[66] The two versions then diverge in the *dénouement* of the fable: according to the first, once she has been able to dispel her jealousy as due to a trivial misunderstanding, a "iucundus nominis error,"[67] Procris is victim of her own relief and of Cephalus's spear, by showing herself unexpectedly and being mistaken for a wild beast. In dying, as she herself observes, she would now have to ironically exhale her spirit upon those very "aurae" (the usage of the plural confirms that she has finally learnt her lesson), which she had once suspected by hearsay, as it were, just because of their name (*nomine*);[68] but Cephalus spares her this posthumous injury by receiving her final breath in his mouth. In the second version, Procris realizes only after being wounded to death that an excess of jealousy had costed her her life, when Cephalus teaches her, all-too late, that hers had indeed been a mere "error nominis."[69]

What undoes Procris is thus equivocation. She mistakes a common noun for a proper name, and pays with her life for what is arguably a mere grammatical error. This moral tale adequately expresses the enlightened conscience of an age, which had been schooled by the *physiologoi* to see in Iris the common noun of a cloud that reflects the sun-rays rather than the proper name of the rainbow.[70] Procris embodies a more archaic linguistic conscience, to which gods only are worthy of a proper name, they only are "namable" (*onomastoi*) in a proper sense.[71] The story makes of her an almost laughable character, a sample out of the psychopathology of everyday life; what she is attempting to do is "simply" to restore "the old meaning"[72] of a name. That she is made to pay with her life for such an attempt, shows *ad abundantiam* that it cannot be a trivial matter. In reality, she knows better than her spouse that names have not lost their mythical powers in spite of Oedipus's solution of the enigmatic incantations (*carmina*)[73] of the Sphynx; her fate comes as the ultimate verification of this insight.

In the second version of the story, the far-removed origin of the tragic events leading to Procris's death is the attempted seduction of Cephalus by the goddess Aurora. Cephalus shows his resolve to break away from the mythical cosmos by choosing Procris over Aurora. But the wounded goddess vows to make him repent of this denial, and submits him a riddle, whose solution he is not able to fathom, differently from his forefather, in time to dispel the prophesized catastrophe.[74]

Cephalus's invocations to the aura sound thus, on the background on this mythical antefact, as unconscious evocations[75] of Aurora's name: Cephalus cannot but recall her divine name when he sings the aura, even if his address is just meant to a natural element. When it resounds anew in Petrarch's poetry, *aura* revives the fading echo of a divine name.[76]

In his exegesis of *Genesis*, Philo wonders why the writer, who shows to know the word *pneuma*, since in the second verse s/he[77] refers to the spirit of God "borne over the waters," is now using a different word to describe the inbreathing of the spirit of God through the nostrils of Adam.[78] Philo explains that the two terms "breath" (*pnoē*) and "spirit" (*pneuma*) are not to be taken as synonymous:

> for 'spirit' is conceived of as connoting strength and vigour and power, while a 'breath' is like a breeze (*aura tis*) or a peaceful and gentle power. The mind that was made after the image and original might be said to partake of spirit, for its reasoning faculty possesses robustness; but the mind that was made out of matter must be said to partake of the light and less substantial air (*aura*), as of some exhalation, such as those that rise from spices: for if they are kept and not burned for incense there is still a sweet perfume from them.[79]

Such a distinction between *pneuma* and *aura* early vanished, probably due to the spiritualization of the letter, which is the enduring linguistic legacy of Paul's apostolate. As a result, *pneuma* overpowered, so to speak, its gentler undercurrent.[80] When we consider, for instance, the way in which Alexander von Humboldt describes his crucial experiment with the *Hauch*, or breath, in his work on Galvanism, and then in a letter to Goethe, we see how divine and human spirit are conflated in one breath, as it were, almost unassumingly.[81] The thigh of a frog, which had been prepared on a glass plate for a series of experiments on animal electricity, was not showing any contraction when linked with armatures of gold or zinc. Only when Humboldt approached by pure chance his mouth to the apparatus, a contraction of the thigh eventually occured, which he recognized as due to the subitaneous volatilization of his breath from the surface [fig. 20]. He evokes the bliss of his discovery in enthusiastic terms:

> Among all the physical experiments I had the pleasure to attempt in the presence of other researchers, I have found no other so astonishing, because of its infinite subtlety, as this one with the breath. The chain of dry metals, gold, zinc, and gold, does not produce any excitation. If one lightly dumps with breath the inferior or the superior

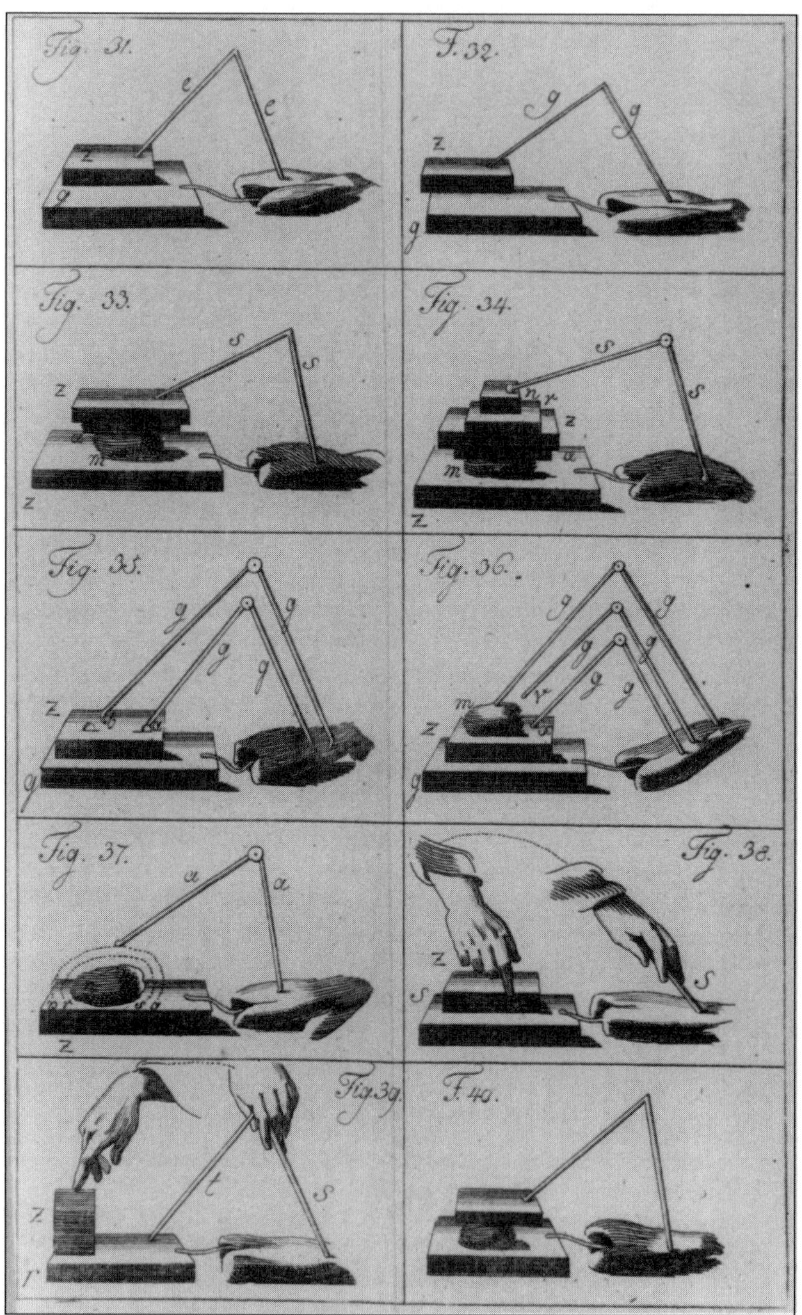

Fig. 20. A. von Humboldt, *Versuch über die gereizte Muskel- und Nervenfaser,* Berlin 1797, vol. 1, pl. IV. Special Collections Research Center, University of Chicago Library.

surface, [. . .] if one lets the gaseous water cover this surface, the
muscle is convulsely shaken. [. . .] If one dries up the surface with
a wollen cloth, the movement disappears again. The experiment looks
like a wonder, in that one at times breathes in life, at time withdraws
the life-giving breath![82]

Goethe, who was not among the bystanders but was informed about the
experiment by Alexander's brother, Wilhelm, was highly impressed by its
results. "I, too, make much of my experiment with the breath (*Auch ich halte
viel auf meinem Versuch mit dem Hauch*)," Humboldt writes in a letter of
reply to the poet, and concludes by a reference to *Genesis*: "It is the principle
of life, the spirit, which hovers over the waters (*Das ist das Lebensprinzip,
der Geist, der über den Wassern schwebt*)."[83]

Humboldt is quoting Luther's translation, which reads: "und der Geist
Gottes schwebet auf dem Wasser." The change of person and preposition in
his wording of this passage, however, suggests that Humboldt meant not just
to quote Luther, but also to pay homage to Goethe by alluding to the title of
one of his most famous poems: the "Song of the spirits over the waters
(*Gesang der Geister über den Wassern*)." Yet Humboldt's allusion is inaccurate in a twofold respect: he singularizes what in the title of Goethe's poem
is a plurality of spirits, and specifies the action of the spirit as a "hovering
(*schweben*)." Neither emendation is warranted. More importantly, Humboldt
conflates and translates in univocal Christian terms the Pagan and Jewish
elements that are inextricably encroached in Goethe's language and give to it
its unique "rainbowy"[84] quality.

The plurality of Goethe's spirits corresponds to the plurality of breezes
blowing over the waters[85] rather than to the unicity of the divine spirit moving
over them. Certainly the plural suggests a choir rather than an *aria*. As far as
the verb Humboldt borrows from Luther, this choice also does not reflect
Goethe's idiosyncratic usage. In a preparatory essay to the translation of the
Bible he projected together with Martin Buber, Franz Rosenzweig pointed out
that Goethe uses the verb *brüten*, instead of Luther's *schweben*, each time he
echoes this specific locus in *Genesis*. Thus Goethe writes in 1784 in the essay
Über den Granit: "when this peak stood still there in the old waters as an
island surrounded by the sea, around it soughed the spirit, which brooded
over the waves (*da dieser Gipfel noch als eine meerumflossene Insel in den
alten Wassern da stand, um sie sauste der Geist, der über den Wogen brütete*)."[86]
The same pattern of substitution occurs in the handful of other occurrences
of the term in Goethe's work. Rosenzweig explains this consistent *variatio*
not as it were a repeated lapsus, but rather as a consequence of the intensive
study of Hebrew Goethe had pursued in his youth, which the fourth book of
Dichtung und Wahrheit documents. Rashi had authoritatively glossed the
Hebrew verb with the French *couver*, and both Buxtorf and the German

translation of the English Bible, which Goethe used, explain the verb as properly referring to the motion of a bird. The same knowledge is reflected by Milton's usage, who also corrects the Authorized Version, which reads "moved," with "brooding on the vast abyss" (*Par. Lost* I.21).[87] Humboldt's allusion is thus, certainly in spite of his best intentions, missing its target, albeit by a narrow measure.

Rosenzweig pursues his exegesis of the passage from the essay *On Granit* by suggesting Herder as Goethe's other possible source. Goethe could have borrowed from Herder's "genial"[88] commentary to *Genesis* in the *Älteste Urkunde des Menschengeschlechts* the other verb he uses, *sausen*, which is implicit[89] in Herder's depiction of the horror of that first night on earth. Rosenzweig concludes his analysis by suggesting that Goethe, in a moment of "bold intuition" of the sense of this passage, was then able to bring together the *Braus* and the *Brüten* in one turn of the sentence, to express how "the tension of that moment of primordial creation still compresses in one that which cannot be united (*die Kräftespannung jenes Augenblicks der Urschöpfung Unvereinbares noch in eines ballt*)."[90]

Rosenzweig interprets this as a peculiar case of "immediate effect of the Jewish Bible on Goethe's language (*unmittelbare Einwirkung der hebräischen Bibel auf Goethes Sprache*)." It is, however, misleading to speak of immediacy apropos of what is, ultimately, a stylistic choice on Goethe's side. If it were immediate, the effect could not fail to occur, like the repeatable outcome of an experiment, as it were. Yet Rosenzweig and Buber themselves eventually opted for a different verb in their own translation, which only partially reflects the results of Rosenzweig's exegesis. The passage reads in their version: "Braus Gottes schwingend über dem Antlitz der Wasser."[91] It is certainly correct to say that Goethe translates directly from the Hebrew, without the mediation of Luther's German, yet he still operates within the boundaries of the German language. The word *brüten*, in other words, does not enrich the lexical spectrum of the German language, even if it is used by the poet in a new *callida iunctura*. Only by compressing together *Unvereinbares*, as Rosenzweig aptly says, in the moment of linguistic creation, a new hue may be added to the rainbow of language. Far from being a mere artifice to give an "air étranger" to one's own style,[92] this is an action of extreme, literally unheard-of violence,[93] as we can sense from another example of Goethe's usage, which may also shed further light on the difference between translation and transliteration.

In his essay on foreign words written shortly before the starting of World War I, Hugo von Hofmannsthal points out an *hapax* in Goethe's work, the unique occurrence of the word *karterieren*, a word Goethe derives by transliterating the Greek verb καρτερεῖν, which means "to be strong, to endure, to resist;" Goethe uses it in reference to "one of his favourite figures," as Hofmannsthal appropriately remarks, the Ottilie of the *Wahlverwandtschaften*:

she must, Goethe says, "karterieren."[94] This may be legitimately termed a specimen of the "immediate effect" of the Greek tongue on the language of the old Goethe. The new word, Hofmannsthal comments, lent him precisely the "nuance" he wanted for this unique usage.[95] Yet we must assume that such a result could only be achieved under the pressure of a creative urgency comparable to the ponderous yet unremarkable pressure of the air, which is responsible for bringing forth such delicate effects as the iridescent colors of the soap-bubbles.[96]

Linguists describe *Fremdwörter*, foreign words, vs. *Lehnwörter*, loan-words, as those words that have acclimatized, but not yet integrated in a foreign language.[97] Adorno singles out Simmel as the foremost example of a writer who managed to escape "the compulsion of identity" the unreflected usage of one's own language necessarily enforces by weaving foreign words in the texture of his own writing; he was thereby capable of giving to his essays "something that eludes official thought—a moment of something inextinguishable, of indelible color;" although, according to Adorno, his gesture was never justified by Simmel in theoretical terms.[98] Adorno himself hints at such a theory when he writes that "*Fremdwörter* are quotations."[99] If we combine this bold generalization with Schuchardt's insight that all words are loan-words,[100] then we have "the brief theory of all corruption of words" Thomas de Quincey wished for:[101] each word is nothing else but the quotation of a foreign word. This implies that, from a theoretical point of view, there are no foreign words: we are always already speaking "with other tongues," since we cannot but use the *glōttai*[102] of which languages are made. Ordinary language is thus already the paradigm of *glossolalia*, in it all words are always already "mottoes,"[103] if only in another language. In another language, namely, words are *mots*. In terms of poetics, this implies that we do not give an *air étranger* to our style by simply grafting on it foreign words; we rather do so by using the words of our own language as if they were themselves foreign.[104] Words used in such a way do not "shine out from the page" because of their "incongruity" with the new context, like "a new patch on an old garment, a *purpureus pannus*,"[105] hence by making appeal to our aesthetic sensibility. When words we have been long familiar with do indeed attract our attention is rather because of the glow that signals the happiness of their choice, as if they were dictated "by a kind of felicity (as a musician that maketh an excellent air in music)."[106]

An extraordinary example of his talent for using German words as if they were *Fremdwörter* occurs in the final piece of Benjamin's *Einbahnstraße*, titled "To the Planetarium." There he refers to the condition preceding and announcing the epileptic seizure as a "Glück," or a bliss.[107] According to the Grimm dictionary, the word *Glück*, which did not originally incline, like most other terms referring to destiny, toward the "good" in preference to the "bad" luck, was used in the sense of a "premonition" or a "presage" up to the sixteenth century, and is found mostly in translations or glossaries, to render Latin terms such as *auspicium, augurium,* and *omen,* especially taken from the birds or the wind.[108] But nowadays, as well as in Benjamin's time, the standard expression in medical terminology to refer to the symptoms of an epileptic seizure is *aura.*[109] The word is a metaphor from the proper sense of aura with which we have become familiar, but in this case we are precisely informed as to the circumstances under which this catachresis could occur. In his treatise *De locis affectis,* Galen narrates how an epileptic he examined at the very beginning of his medical practice would describe his condition as originating in the lower leg and from there climbing "upwards in a straight line through the thigh and further through the flank and side to the neck and as far as the head; but as soon as it had touched the latter he was no longer able to follow." As he could not explain what exactly was rising up to his head, a bystanding youth, who also suffered of the illness, was able to supply him with a fitting description: "it was like a cold breeze (*oion auran tina psychran*)."[110] From this moment on, thanks to the linguistic creation of a sufferer, who was certainly experiencing no bliss, but was rather "forced to coin" a name for his pain,[111] physicians could identify this prognostic of the coming seizure of the disease, whose status had meanwhile decayed from "sacred" to "caducous:" *passio caduca.*[112] We can now restore Benjamin's text: "In the nights of annihilation of the last war the frame of mankind was shaken by a feeling that resembled the *aura* of the epileptic," Benjamin writes, "And the revolts that followed it were the first attempt to bring the new body under its control."[113] But a *hapax* such as this cannot be explained by purely philological means.[114] The question we have still to ask ourselves is: why would Benjamin translate, knowingly or unknowingly, *aura* with *Glück*?

A text written in the middle of those very convulsions he evokes in "To the Planetarium" may help us understand the reasons of this substitution. In the so-called "Theologico-Political Fragment" Benjamin asserts "with all possible clarity,"[115] in consonance with Ernst Bloch's *Geist der Utopie,* that "the order of the profane cannot be constructed in terms of the idea of the Divine Kingdom."[116] However, as the immediately following sentence states with no less clarity,[117] there is an "aim" (*Ziel*), to which the order of the profane ought to turn itself without fearing to find in it its "end" (*Ende*), but rather its "downfall" (*Untergang*) and its subsequent "*restitutio in integrum*"— namely, "the idea of happiness" (*die Idee des Glücks*).[118]

Glück is thus both the symptom and the therapy of our mortality. Happiness is a promise of immortality: *le bonheur,* even if it were just *une bonne heure,*[119] is nonetheless *une promise d'immortalité.*

Chapter 5

Nemesis and *Aphanisis*

Quidquid latet apparebit,
Nil inultum remanebit.

Dies irae

Nemesis's name may be today a common noun, or even sound like a "commonplace,"[1] yet its mention still evokes at least part of the awe the Greeks felt as they invoked it: for we all dread our nemesis. As it points toward an almost individual power, the name thus serves as a reminder of its divine origin. *Our nemesis* is a personal chastiser, meant just for us, a personification that still reminds us of our mortality or human limitation, even in the absence of a religious belief, a messenger (an angel: *angelos Dikēs*)[2] of Justice, if not Justice herself. Yet Nemesis was a common noun already to the ancients, at least by the time of Aristotle, who has left us the legacy of a definition that emphasizes its distance from a religious belief: between the two extremes of envy and maliciousness[3] is

> what the ancients call Righteous Indignation (*Nemesis*)—feeling pain
> at undeserved adversities and prosperities and pleasure at those de-
> served; hence the idea that Nemesis is a deity.[4]

Nemesis was indeed a deity of ancient lineage in the Greek pantheon. We find her first mentioned by Hesiod in his *Theogony*, as a daughter of the Night—a genealogy that hints at a kinship with the chthonic world from which the Olympian religion emerged;[5] but the other Hesiodean poem, *Works and Days*, already laments her departure from the human world and her rejoicing of the "deathless gods," along with Shame (*Aidōs*);[6] though her very disappearance is here interpreted as a step toward her apotheosis.

In a more ominous, less flattering way, the very memory of Nemesis threatened to be obliterated by the syncretism of the medieval mythographers: chiefly remembered as a synonym of Fortune, or explained as a moral quality, such as *benignitas*, by St. Bonaventure,[7] Nemesis was indeed a dead divinity, her name dead letter by the time of her rediscovery in the Renaissance. Poetry and art then rescued her from the long oblivion by recalling and representing her. The Italian humanist and poet Angelo Poliziano deserves full credit for initiating such a revival. Poliziano's poem *Manto* (1482) is unquestionably "the first modern attempt to deal verbally with Nemesis at any length;"[8] in turn, it inspired the first modern attempt to deal figuratively with her, Albrecht Dürer's full-length portrayal [fig. 21].[9] The German artist depicted the goddess in an engraving that was long known as *Das grosse Glück* (*The Large Fortune*), in spite of his own identification of the subject.[10] The failed recognition only confirms that her reappearance had hardly been anticipated. The first part of Poliziano's poem provided Dürer with an almost hallucinatory description of the goddess, walking "afloat, floating in empty air;"[11] after a detailed overview of her apparel, Poliziano then proceeded to remind his audience of her moral prerogatives:

> She subdues extravagant hopes; she threatens the proud with dangers; to her is given power to crush the arrogant minds and triumphs of men and to confound their ambitious plans. The ancients called her Nemesis [. . .] Exchanging high and low, she mixes and tempers our actions by turns, and she is borne hither and thither by the whirling motion of the winds.

The lofty description is followed by a sort of historico-philosophical digression on Nemesis's role and intervention in world affairs, which serves to prepare the transition to the main topic of the poem, a celebration of Virgil's poetry. After punishing the arrogance of the Persians, it is now the turn of the Greeks to suffer Nemesis's wrath at the hands of the Romans:[12]

> She had seen how you, Greece, swollen from the conquest of the Persians, carried your victorious arms to the eastern part of the globe; she had seen how you rode high, proud of muse-inspired song and eloquence, and how you bragged, raised your upturned head to the stars and believed yourself to be equal to the gods. But soon, detesting noxious haughtiness, she forced you to wear the yoke upon your neck and subjected you, vanquished, to the arms of the Romans.

Poliziano hints here at a well-known anecdote, according to which the sculptor Agorakritos, after the Greek victory over the Persians, turned a block of marble the enemies meant to use in order to commemorate *their* victory into

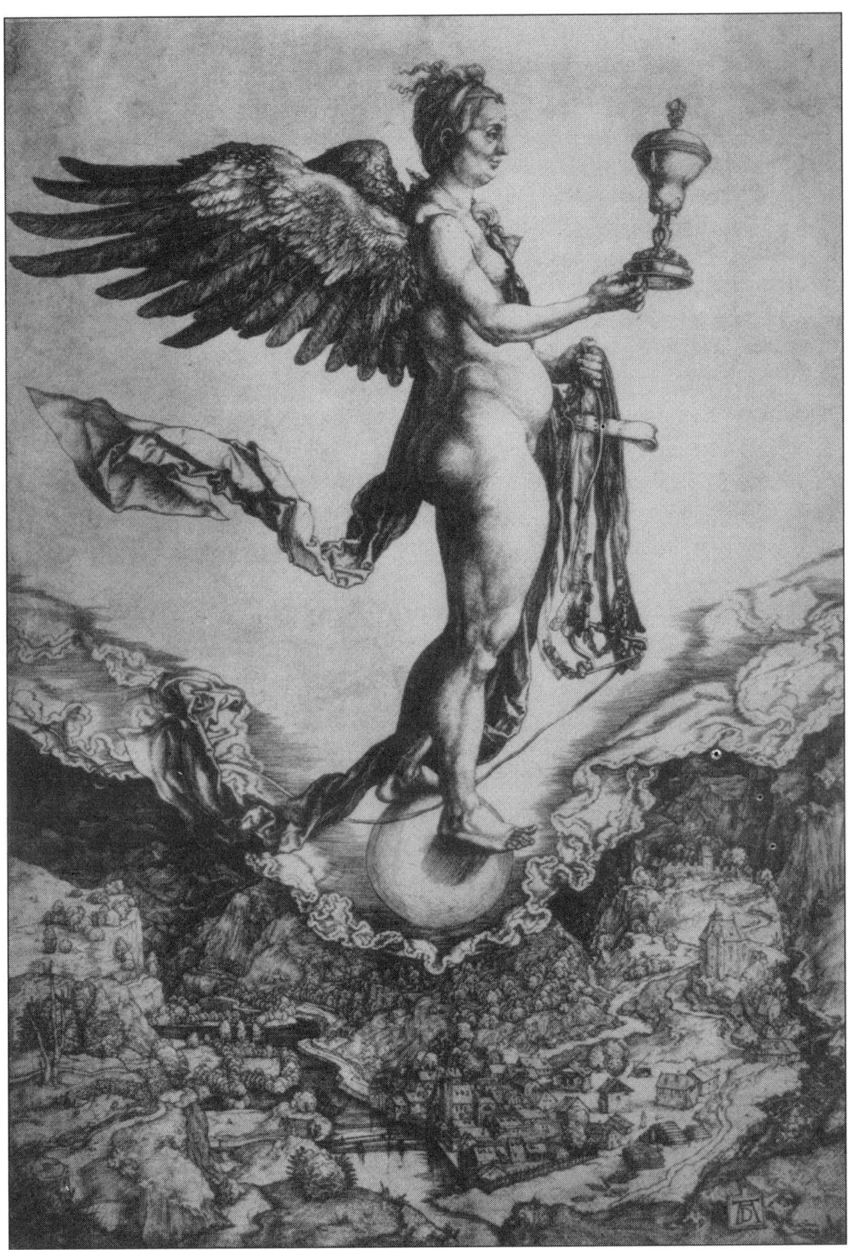

Fig. 21. Albrecht Dürer, *Nemesis,* engraving, ca. 1500–1502. Reproduced from *Albrecht Dürers sämtliche Kupferstiche in Grösse der Originale* (Leipzig: Hendel 1928). McCormick Library of Special Collections, Northwestern University Library.

a statue of the goddess of revenge—Nike ironically metamorphosed into Nemesis, a truly Kafkaesque conceit! And the reference to Kafka is not forced here, if we consider the conflation, which strikes at once K. as contradictory, of Justice and Victory in one of Titorelli's painting: how, he wonders, can Justice be winged *and* the scales she holds be balanced?[13]

Poliziano's sources are Pausanias and, most directly, two epigrams from the Greek anthology. The first, which Ausonius rendered into Latin and was probably Poliziano's main inspiration, lets the goddess herself tell the tale:

> As a stone the Persians once brought me here to be a trophy of war; now am I Nemesis. And even as I stand here a trophy of Greek victory, so as Nemesis I requite the idly boasting Persians;[14]

the second, more subtly, lends words to the stone itself, which avows the ambiguity of its claim to worship:

> I am a white stone which the Median sculptor quarried with his stonecutter's tools from the mountain where rocks grow again, and he bore me across the sea to make of me images, tokens of victory over the Athenians [. . .] but now I am Victory to the Athenians, Nemesis to the Assyrians.[15]

Nemesis was represented as holding a cubite-rule and a bridle, and her motto was "nothing beyond due mesure (*mēden yper to metron*)."[16] It is not far-fetched, then, though certainly idiosyncratic, that the Italian poet chose to put his eulogy of Virgil under the aegis of a goddess who particularly disliked lack of measure in utterance.[17] As he chooses to write in Latin hexameters, which often echo those of his model, Poliziano builds a monument to Virgil, the poet who rivalled the achievements of Greek poetry in Latin verses; for not even the "primacy of eloquence" is left to the Greeks, after the loss of their political power and freedom, as they "spontaneously" transfer to the new rulers the crown of poetry (*sponte tibi virides transcribens Graecia palmas*).[18]

In choosing the verb *transcribere*, which means chiefly "to transcribe," in the sense of copying, Politian was probably thinking of a particular occurrence of the verb in Pliny's *Natural History*, the great encyclopedia of classical antiquity, a work highly appreciated by the humanists. In the preface to his work, avowedly a compilation, Pliny prides himself of having openly acknowledged his sources, differently from many of his contemporaries:

> when collating authorities, I have found that the most professedly reliable and modern writers have copied the old authors word for word, without acknowledgement (*veteres transcriptos ad verbum, neque nominatos*),

though certainly "not in that valorous spirit of Virgil, for the purpose of rivalry" (*non illa Vergiliana virtute, ut certarent*).[19] Virgil did not imitate but rather emulated (a *topos* that Petrarch and his fellow-humanists were keen to reinforce), and could thus legitimately challenge the supremacy of the Greek epic poets, as Cicero had already shown to be equal to their best orators.

In spite of the Greek defeat, however, the name of the goddess defies translation. Even if "there is at Rome an image of the goddess on the Capitol," Pliny wonders, yet "she has no Latin name."[20] A statue can be transferred, or copied, but not a name. A name cannot be translated, only recalled. In spite of Poliziano's celebration of the *topos* of the *translatio imperii*, his mention of the name of the goddess unwittingly celebrates her as a figure of transliteration. The goddess kept her name, though in exile. If the *translatio imperii* is, truly, the empire of translation, then transliteration is the true revenge of the defeated: "Greece, though captive, took her savage victor captive, and brought the arts into rustic Latium" (*Graecia capta ferum victorem cepit et artis intulit agresti Latio*).[21]

In the Homeric poems Nemesis does not occur in person yet, though her name does. In the third canto of the *Iliad*, when Helen first appears on the walls of Troy, her appearance in the text is accompanied by *nemesis* in a negative formula that is not unique in Homer, though certainly not formulaic in this particular instance:

> We cannot rage (*ou nemesis*) at her, it is no wonder
> that Trojans and Akhaians under arms
> should for so long have borne the pains of war
> for one like this.[22]

The unwillingness of the Trojan elders to blame Helen for their predicament is a direct consequence of their recognition of her as a goddess: "Unearthliness. A goddess is the woman to look at."[23] To blame her would not just be wrong, or naive, to blame her would be blasphemous, for her appearance casts the viewers under the spell of the very goddess they would otherwise invoke as her chastiser. The "ghastly" (*ainos*) resemblance between Helen and Nemesis is indeed not uncanny, if we consider that Nemesis was Helen's mother in the epic cycle to which the Homeric poems belong. The prehistory of the war was narrated in a series of poems, among which is the *Cypria*. A fragment of the poem records the myth, according to which Nemesis gave birth to Helen after being pursued by Zeus through a series of metamorphosis: in the shape of a swan, he finally succeeds in overpowering the goose-like Nemesis, and fecondates the goddess. Before succumbing, Nemesis tries to escape the incestuous union with the "father" Zeus, as "her heart was vexed by shame and indignation (*aidōi kai nemesei*)."[24] Thus, even if she apparently yields to a superior power, the goddess ultimately remains faithful to her own nature and reclaims her freedom: Nemesis can only be haunted by herself, as it were, for *aidōs* and *nemesis* are still an hendyadis to the poet of the *Cypria*, as they were to Hesiod, who lamented, as we have seen, their simultaneous abandonment of the human world.[25]

We can measure how the profile of the goddess changes if we move from this early family novel to a late epigonic saga, the last great example of Greek epic poetry, Nonnos's *Dionysiaca*. In the lengthy poem, a full-fledged catalogue of Dionysius's feats and misdeeds, Nemesis is introduced as a prosecutor, eager to punish any trespass: instead of being chased after, she now flies down to earth and hunts down the culprit with the help of her heraldic companion, the griffin, "a bird of vengeance" (Shelley's *alastor*).[26] Nemesis's victim is here another personification we are already familiar with: Aura. Nemesis is invoked by Artemis in order to punish the arrogance of "the Windmaid (*Aurē*)," whose "name was like her doings:" she "could run most swiftly, keeping pace with the highland winds (*aurai*)" (vv. 256–257). A companion and devotee of Artemis, but contemptuous of Aphrodite, Aura dares to compare her body, "like a boy's, and her step swifter than Zephyros" (362–363), to that of her mistress: Aura unashamedly "scanned the holy frame of the virgin who may not be seen" (341–343) as she was bathing naked, and loudly questioned the legitimacy of Artemis's claim to "the name of a virgin maid" (351), because of the opulence of her forms. Wounded, the goddess applies for the intervention of Nemesis to check "the faults of [Aura's] uncontrolled tongue" (432). As a punishment for her arrogance, worthy of Dante's *contrapasso*, Aura, "the champion of chastity (*philoparthenos*)" (430), is violated by Dionysus. What comes around, goes around: the wheel also figures prominently among Nemesis's attributes. Her decree re-establishes the broken balance and placates the irate goddess: "Aura the maid of the hunt has reproached your virginity, and she shall be a virgin no longer" (445–446). Dionysius takes advantage of her sleep to violate her, and the outraged Aura realizes her loss when it is already too late: she gives birth to twins, not without having to endure Artemis's mockeries, who ridicules her as a "virgin mother" (859). Unaware of the plot spinned by Nemesis, Aura can only blame her own eponyms, the *aurai*, as authors of her disgrace: "I was wooed by the breezes, and I saw no mortal bed. Winds my name-sakes (*eponymoi aurai*) came down to the marriage of the Windmaid" (893–894). Differently from Procris's, though, Aura's parting words are unfair. After killing one of the newborn, and overwhelmed by shame for the change of her name from virgin to bride,[27] Aura throws herself into a river and is mercifully transformed into a fountain by Zeus.

As is evident even from such a schematic retelling of the myth, Nonnos splits in two the figure of Nemesis that we met in the *Cypria* haunted by her own self: here Nemesis is turned into an external instance, Aura her victim, who dearly pays for what, once again, might appear to us just a misdemeanour. But we also find traits of the original Nemesis in the two wounded goddesses, who both resent Aura's unchecked speech: we can recognize her in both Artemis, persecuting Aura because of her blasphemy, and Aphrodite, who was shunned by Aura[28] and, though not personally involved in the plot, must now rejoice at her punishment. Such a feature of the saga cannot but remind us of Aphrodite's rage against Psyche.[29] Nemesis persecutes Aura, as Aphrodite Psyche, whose only guilt is to be called, because of her beauty, by the name of the goddess. It is the profanation of her name that awakens Aphrodite's wrath: "my name, built up in heaven, is profaned by the mean things of earth!"[30] The greatest danger for the divinity is homonymy, the removal of the taboo that does not allow to invoke in vain a divine name. Once a divine name is employed by the mortals in ambiguous way, it loses the dignity of a proper name and decays to the status of a common noun. Nemesis is the instance meant to redress such *lapsus linguae*. On the other hand, the oblivion of their names represents an equally mortal injury to the gods. Yet, even when neglected, Venus warns Cupid, "a goddess grows in power."[31]

According to Pausanias's description, the statue of the goddess in Rhamnous had no wings, though, he continues, "later artists, convinced that the goddess manifests itself most as a consequence of love, give wings to Nemesis as they do to Eros."[32] Her connection with Aphrodite is confirmed by another anecdote Pliny relates about the origin of the Rhamnousian Nemesis. According to this version, Agorakritos, disappointed after his statue of a Venus had been unfairly rejected by the Athenians, offered it to Rhamnous provided it would be renamed as Nemesis.[33] Human language is necessarily guilty of metonymy, but any misnomer is an outrage to Nemesis: Nemesis's name is the name's nemesis.

Nemesis presides over an inconspicuous part of the body, consonant to her nature and, one might say, to her anonymity in the Latin language, which Pliny points out once again with incredulity:

> behind the right ear is the seat of Nemesis (a goddess that even on the Capitol has not found a Latin name), and to it we apply the third finger after touching our mouths, the mouth being the place where we locate pardon from the gods for our utterances.[34]

This usage sanctions the role of Nemesis as "the foe of our tongues (*glōssēs antipalon*)."[35] Impious statements bring about divine retribution, and they cannot elude Nemesis's recollection, for she literally writes them down on the surface consecrated to her.[36] Following the example of the goddess with the classificatory zest of the naturalist, Carl von Linné, the famous eighteenth-century botanist, gathered during the course of his life a collection of such statements, complete with the ensuing punishments, under the heading of *Nemesis Divina*. The result is a cento of quotes meant to prove that one cannot escape divine retribution, and that we are ultimately to blame for our own misfortunes. Linné's Nemesis, though, in spite of the epithet, is certainly closer to an abstract concept of compensation, or retaliation,[37] than to the Greek divinity with whom we have become acquainted.

In introducing the goddess to his contemporaries and, more importantly, lending her name to the vocabulary of German Romanticism, Herder was careful to differentiate his more refined understanding from Linné's "vulgar" view of the goddess.[38] His Nemesis is not meant to be a divinity that terrifies, but rather one worthy of being loved: for "her frightening name has become frightening only because of misunderstanding."[39] Herder's 1786 essay is precisely meant to rescue Nemesis's name from its bad renown, if not from oblivion, and to surround it again with its lost aura. In order to do so, he combines the mythological data we have discussed so far into a coherent didactic tale (*ein lehrendes Sinnbild* is the subtitle of the essay). Nemesis becomes thus a moralized Venus and a sister of Shame,[40] to such an extent she may be also called beautiful:[41] "she who was once Venus is now transformed in a virtuous, chaste goddess."[42]

Goethe was probably thinking of this metamorphosis when he returned the manuscript of the essay to his friend's wife, accompanied by the lines "Youthfully she comes from the sky" (*Jugendlich kommt sie vom Himmel*).[43] The quatrain epitomizes Goethe's reading: the goddess, who descends bare from the heavens and stands unveiled in front of priests and sages, is covered with a transparent veil of sacrificial smoke by (an unnamed) Herder, his eyes downcast, and is thus made bearable to human sight. The lines, though flattering, expose Nemesis's metamorphosis as the performance of a speculative legerdemain. Yet another aspect of Herder's analysis had to be troubling to Goethe: his subjection of hope to the jurisdiction of Nemesis. In Herder's view, Nemesis's rule implies compliance with an internal measure, not merely an external limit. Only the "internal Nemesis of his thoughts (*innere Nemesis seiner Gedanken*)" may remind the one blessed by fortune of his limits: "he must learn to bridle himself, even if Hope puts wings to his steps."[44] The sentence had to sound to Goethe like a warning directed personally at him, the Fortune's darling who read the essay while in Rome, and had to wonder "how someone could have written something like that, without having been to Italy"![45] We have just to recall the final words of *Dichtung und Wahreit*, the famous quote from *Egmont*, in order to understand how Goethe would take Herder's invitation to moderation. His unwillingness or inability to let the bridles be handled by anybody else is made painfully clear:

> The coursers of time, lashed, as it were, by invisible spirits, hurry on the light car of our destiny; and all that we can do is in cool self-possession to hold the reins with a firm hand, and to guide the

wheels, now to the left, now to the right, avoiding a stone here, or a precipice there. Whither it is hurrying, who can tell? One hardly remembers whence it came.[46]

Even if Herder's essay had the merit of first drawing Goethe's attention onto Nemesis, it is to another essay devoted to the goddess that we owe one of the most enigmatic texts of Goethe's old age, the five stanzas gathered under the title *Urworte. Orphisch.* In 1817 Goethe read an essay on "Tyche and Nemesis" by the Danish philologist and antiquarian Georg Zoega, meant in part as a corrective to Herder's moralization of the goddess.[47] Herder's interpretation of Nemesis as "the goddess of measure and restraint" is dismissed as too limited by Zoega, whose far more ambitious goal is to present Nemesis as "the source of all justice, the legislator of the universe, the mother of destiny."[48] Zoega takes the polynymy of the ancient gods as his starting point, and makes his task to sort out the meaning of such "hieroglyphs."[49] He shows that the two divinities, which are united under his title, were already confused and mistaken by the ancients, but *pour cause*: they are both representatives of "the order of things that are independent from man."[50] As a result of his analysis, Tyche is then identified as pure chance, the power of fortune that rules each individual's destiny from without, as opposed to the inner steering principle, the *daimōn*,[51] and to the overarching influence of Nemesis, "the ultimate final cause, which imparts to each of the apparent causes their measure of efficacy and strength."[52] Her action is thus consonant with the etymology of her name, which Zoega, like most scholars, derives from the verb *nemein*, whose meaning is precisely "to impart, to distribute."[53] But the name, according to him, and here his interpretation is more idiosyncratic, was only at a later stage bestowed upon a divinity the Greeks had imported from Egypt: Adrastea, personification of the Night. Zoega is thus also able to explain Nemesis's connection with Aphrodite on this ground: the nightly starry sky was rightly regarded as the most beautiful goddess.[54]

In this context Zoega quoted a passage from Macrobius, which ostensibly inspired Goethe's composition of the *Urworte*: "According to the Egyptians, the gods who attend a man's birth are four: Δαίμων, Τύχη, Ἔρως, Ἀνάγκη."[55] Goethe's series of poems is meant as a meditation on this anecdote, and his own later prose commentary (1820) further elaborates the programmatic meaning of the cycle. To the four sacred words (Macrobius calls them *ieroi logoi*), *Daimōn, Tychē, Erōs, Anankē,* Goethe adds a fifth one, *Elpis.* Goethe, though, does not deem necessary to explain his choice of Elpis over Nemesis, seemingly a more obvious candidate, as she is the protagonist of Zoega's essay; moreover, the latter text had to remind him of Herder's former handling of the matter.[56]

In the literary and iconographic tradition there is a clear antagonism between Nemesis and Elpis.[57] Poliziano hints at it as he writes: "She subdues extravagant hopes (*spes immodicas*)," a line that in turn echoes an epigram of the Greek Anthology: "I counter-balance vain hopes."[58] A funerary stele provides the appropriate locus for the somber inscription: "some Nemesis (*tis Nemesis*) has overthrown your hope,"[59] which testifies to an advanced grammaticalisation[60] of the sacred name. The visual counterpoint of the two goddesses, which is best illustrated by two emblems from Alciati's collection [fig. 22], [61] is verbalized in another epigram Herder used as epigraph to his essay:

> *Nemesis und die Hoffnung verehr' ich auf Einem Altare;*
> *"Hoffe!" winket mir Die; Jene: "Doch nimmer zu viel!"*

> I worship Nemesis and Hope on one altar; "Hope!",
> one winks at me; but the other: "Never too much!"[62]

Such a warning was wasted on Goethe [fig. 23].[63] His commentary to the *Urworte* does not extend to the last stanza, as he devolves to the readers their interpretive freedom. Making use of such a license, I suggest that we read the *Urworte* as a belated reply to Herder's essay, and the last stanza as a celebration of the emancipation of Elpis from Nemesis.[64] Elpis is described as "unbridled" (*ungezügelt*), precisely to mark such an emancipation. Endowed with wings "with her, through her," we can also fly: no longer a descent to earth of the goddess, the flight is rather our ascent, our elevation through the goddess. With just a wingbeat, we may leave behind us ages: "*Ein Flügelschlag—und hinter uns Äonen.*"[65]

Though unnnamed, Nemesis triumphs nonetheless in the celebration of her synonyms. For Elpis and Nemesis are two names of a same *numen*. Goethe had to be aware of a passage from Dio Chrysostomos, which Zoega also quoted in the opening lines of his essay:

> Fortune (*Tyche⁻*) has been given many names among men. Her impartiality (*to ison*) has been named Retributive Justice (*Nemesis*); her obscurity (*to ade⁻lon*), Hope (*Elpis*); her inevitability, Fate (*Moira*); her righteousness, Law (*Themis*)—truly a deity of many names and many ways.[66]

The conclusion of Zoega's essay had certainly to gratify Goethe more than Herder's ominous epigram, as it suggests a possible identification of Nemesis with the *agathē tychē*, the good Fortune, or, *das gute Glück*.[67]

Macrobius offers us also another interpretation of Nemesis in his *Saturnalia*:

> Nemesis, which we worship to keep us from pride, is none other
> than that power of the sun whose nature is to make dark the things
> that are bright and withdraw them from our sight and to give light
> to things that are in darkness and bring them before our eyes.[68]

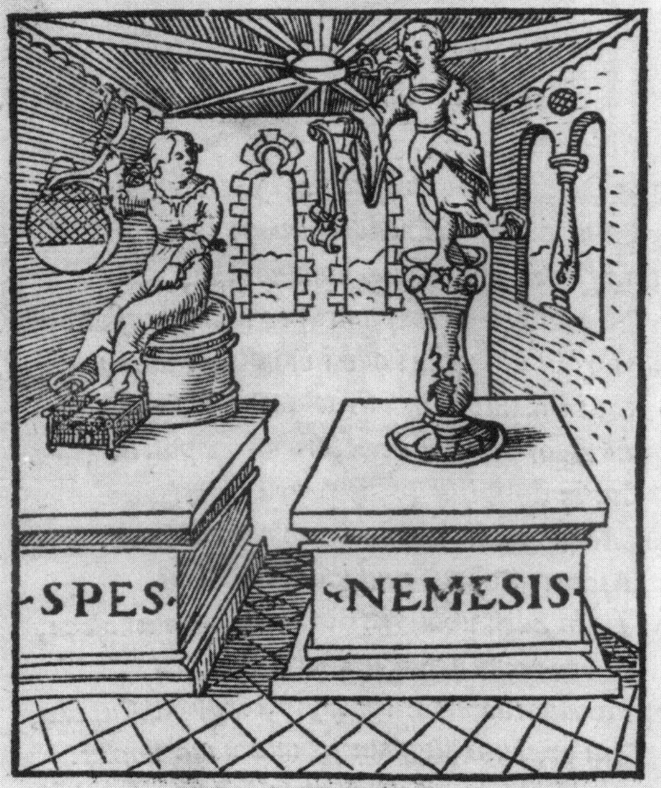

Fig. 22. Andrea Alciati, *Emblematum libellus*, Paris 1542. McCormick Library of Special Collections, Northwestern University Library.

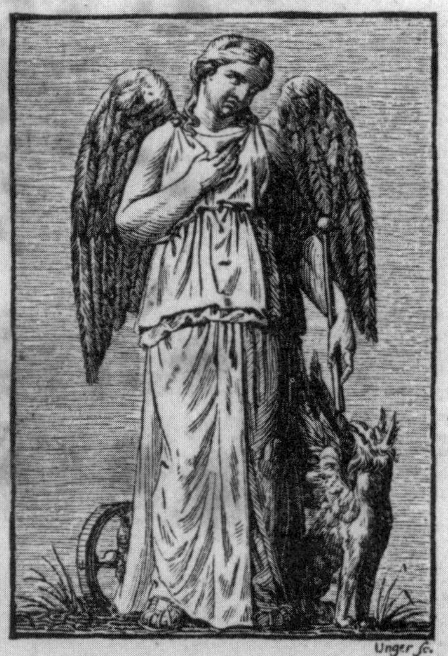

Fig. 23. Johann Wolfgang von Goethe, *Neue Schriften,* vol. 7, frontispiece, Berlin 1800. McCormick Library of Special Collections, Northwestern University Library.

Though unacknowledged in this particular instance, but even more conspicuous by his absence,—as he is one of the authors Schelling preferentially relies upon—Macrobius probably inspired Schelling's own interpretation of Nemesis in his *Philosophy of Mythology*. Schelling gives to Nemesis a very prominent role in his lectures as she embodies, in his presentation of the mythological world order, the "world law" that precedes and presides over the very origin of human freedom. In Schelling's interpretation of Nemesis, she is

> nothing else but the power of that highest world law that brings everything in motion, does not want that anything remain hidden, drives out everything hidden, and as it were morally forces it to show itself.[69]

Almost molded word for word on Macrobius's neo-Platonic interpretation of the goddess,[70] Schelling's definition is also certainly reminiscent of the evangelical prophecy: "there is nothing covered, that shall not be revealed; and hid, that shall not be known;"[71] even more so, of its apocalyptic echo in the Joachimitic hymn *Dies Irae*: "quidquid latet apparebit,/nil inultum remanebit."[72] But rather than the prefiguration or the consummation of a final judgement over man, Schelling's Nemesis is the messenger of human justice. She announces and makes possible its advent, as she forces man to come out of his self and experience the limits of his nature. In a passage reminiscent of Pico's *Oration on the Dignity of Man*, Schelling presents humanity as caught between the two extremes of actuality and potentiality, with a dawning consciousness of its freedom: Nemesis forces man to acknowledge his ambivalent nature (*natura anceps*),[73] and to decide that he must abandon the absurd claim to be *already* equal to God, though his creature; rather, he must legitimate his aspiration to a higher self by recreating himself as equal to his image. Nemesis can awake man's aspiration to freedom precisely because she is the mythological legislator, as it were, without which freedom would be a void concept. Nemesis is the name of the law in a literal sense, as Schelling points out that *nemesis* and *nomos* are etymologically related.[74] Nemesis is the imposition of the law, what Hölderlin would have called the immediate,[75] ruling before and above both mortals and immortals; the law she posits opens up the possibility of transgression. But even according to the Christian view, as expressed by "the deepest of the apostles," the law is the cause of the crime, since "the God-given law [. . .] provoked the sin, namely, the deviation from the original being."[76] Like the expulsion from Eden, then, Nemesis is an automatic[77] retribution, which is unwillingly triggered by gods and men alike, but she is not a moral power: Nemesis is a justified, at most justifiable, not a just indignation—Pindar dares to call her "over-just (*yperdikon*)."[78] The concept of a natural *and* just retribution is already an epigonal one, which, if at all, applies to *nomos*, but not to *nemesis*. The concept of justice has not to be identified with Nemesis.

Schelling takes Nemesis as a power before the law, before justice, which cannot be judged according to moral standards nor be moralized. The law is herself unjust, that is the mystery the Olympian religion wants to hide, "the violence of that uncanny principle that ruled in the earlier religions (*die Gewalt jenes unheimlichen Princips, das in den früheren Religionen herrschte.*)"[79] Homer, traditionally viewed as the creator of Greek polytheism,[80] does hide again that which Nemesis brings to light, poetry "covers with flowers" the "abyss" over which the Homeric world is built,[81] veils that which cannot be contemplated: "Homer, namely, the Homeric polytheism, is only resting upon this forgetfulness of the mystical."[82] In other words, upon the oblivion of Nemesis's name. *Nomos* replaces *Nemesis* as the name of the law in the human cosmos. The order of revenge replaces the natural order.[83] Such is the human order poetry makes bearable, but only at the price of replacing the name of god by the names of the gods. The proper name of god becomes the common noun of the gods. Nemesis's name is the forgotten name of god.

It is in this context that Schelling provides the (parenthetical) definition of the German term *unheimlich* that so powerfully drew Freud's attention,[84] and reinforced his own interpretation of the term: "'Unheimlich' is the name for everything that ought to have remained [. . .] secret and hidden but has come to light."[85] As we have seen, *unheimlich* is above all the name of Nemesis *qua* name of god. The name of god must be forgotten, or repressed, that which means: translated. Yet the name keeps returning out of oblivion in its transliterated form.

Freud was probably unaware of the theoretical context of Schelling's definition, as he gathered it out of a lexicon. Rather than being inspired by Schelling's dialectical analysis, Freud's interpretation is more obviously indebted to another, though parodistic, philosophy of mythology, the one exposed by Heine in his *The Gods in Exile*. "The 'double' has become a thing of terror," Freud thus renders the gist of Heine's argument, "just as, after the collapse of their religion, the gods turned into demons."[86] Heine had ironically suggested that the Greek divinities might have survived as the demons of the Christian world, their fate being exile rather than extinction. The Christian demonization of the word *daimōn* itself may provide the best evidence in support of such an hypothesis, while its very survival certainly bears the most eloquent testimony to the resilience of transliterated words. *Daimōn* had indeed puzzled its Latin interpreters, starting with Cicero, and continues to embarass modern translators.[87] The most successful attempt at a translation is Apuleius's *genius*, which he proposed in a still tentative way in his treatise on Socrates's sign: "in our language we can call him Genius, as I translate, I do not know whether well, certainly at my own risk (*Eum nostra lingua, ut ego interpretor, haud sciam an bono, certe quidem meo periculo poteris Genium uocare*)."[88] In Apuleius's typology, the translation is meant to refer more precisely to the good demon, a variety that is clearly not contemplated by Christian theorists. The Fathers, however, will be able to recuperate the unorthodox idea of a personal god presiding over each individual's destiny thanks to Origen's ingenious hypothesis: that baptism has the power to transform the evil demon, under whose tutelage we are born, into a good angel.[89] The good demon could thus become the guardian angel, the *genius* a secularized term already by the time Xylander used it in his translation of Plutarch's treatise *De genio Socratis*.[90] The most nefarious consequence of such a translation is that it obliterates the relationship that exists in Greek between the *daimōn* and happiness, *eudaimonia*. One of the highlights of the agon between Aeschylus and Euripides in Aristophanes's *Frogs* is Aeschylus's criticism of the opening lines of his antagonist's *Antigone*: "A happy (*eudaimōn*) man was Oedipus at first,/ Then he became the wretchedest of men," to which Aeschylus retorts that Oedipus was "unhappy by nature" (*kakodaimōn physei*), for, "not yet born nor yet conceived," Apollo foretold he would be his father's murderer: how could he be "a happy man at first"? Born wretched, he never ceased (*epausato*) to be such.[91] The two poets bespeak here two clearly alternative views of the relationship between happiness and character. Aeschylus is still aware of the presence of a divine name in that which Euripides already regards as just a qualifier. A less secularized attitude, though already epigonal if compared to Aeschylus's, comes to the fore in the final pages of Plato's *Timaeus*, where Plato calls *eudaimōn* him who "is for ever tending his divine part (*to theion*) and duly magnifying that daemon who dwells along with him (*ton daimōna xunoikon en autōi*)"[92]—a passage that, incidentally, also proves that *daimōn* had already almost become synonymical of "divine" by the time of Plato.[93] *Daimōn*, on the other hand, cannot be taken as a synonym of *theion*: its interpretation as the Apportioner, if the derivation from *daiōmai* is correct,[94] would rather assign it to the sphere of *nemesis*.[95] Like Nemesis, *daimōn* is a power that precedes the humanization of the gods and their polynymy:

> Daimon is the veiled countenance of divine activity. There is no image of a daimon, and there is no cult. Daimon is thus the necessary complement to the Homeric view of the gods as individuals with personal characteristics, it covers that embarassing remainder which eludes characterization and naming.[96]

It is obvious, however, that only the modern interpreter is liable to such an embarassment. The *daimōn* is patently not anonymous, it only eludes naming to the extent that we keep forgetting its name. The name of god is the name that constantly eludes us.

The *daimōn* is nothing else but the double of god,—an insight both the Pagan and the Judeo-Christian tradition seem to ultimately share. If, according to Schelling's most extreme formulation, "the gods *are* the hidden god,"[97] then the demon is the uncanny double whose appearance necessarily preludes to the return and recognition of the true god,—as Ulysses's disguise is a necessary guile in order to avenge the suitors' attempted usurpation of his throne. The *revenant* heralds his own return in the demonic[98] form of the foreigner. But whether we read dialectically, like Schelling, the development of religion as moving toward a form of purified monotheism, or parodistically, like Heine, toward a bourgeois twilight, we must nevertheless conclude that the gods keep returning as demons, namely, unrecognizable in their travesty as common nouns,—including *daimōn* itself.

On the other hand, as we have seen, Goethe tried to rescue the name from its Christianized spelling by restoring its original Greek in the *Urworte* and, even more so, by employing the German transliteration *dämonisch* as a key term in his autobiography. The uncanny effect the return of a repressed name produces is here emphasized by Goethe's choice of the word in order to name something he vouches unnameable:

> He thought he could detect in nature—both animate and inanimate, with soul or without soul—something which manifests itself only in contradictions, and which, therefore, could not be comprehended under any idea, still less under one word. It was not godlike, for it seemed unreasonable; not human, for it had no understanding; nor devilish, for it was beneficent; nor angelic, for it often betrayed a malicious pleasure. [. . .] To this principle, which seemed to come in between all other principles to separate them, and yet to link them together, I gave the name of Demonic (*Dieses Wesen [. . .] nannte ich dämonisch*), after the example of the ancients.[99]

Goethe introduces the term as he describes the process that lead to the composition of *Egmont*: at a particularly critical time in his life, he "tried to screen *him*self from this fearful principle, by taking refuge, according to *his* usual habits, in an imaginary creation." In creating Egmont's character, however, Goethe's customary mechanism of defense was taking on nothing less than "the most fearful manifestation of the demonic" itself, its embodiement in an "individual character."[100]

If the character Egmont thus becomes his screen in the hindsight of the autobiography, Goethe's extended screen-memory, it is nonetheless obvious that the true screen is the very word *dämonisch* the poet feels now free to pronounce. The recovery of the transliterated name is meant to lend him control and mastery over the fearful forces he has evoked. But the author of the *Faust* was certainly not unaware of the dangers involved in such an operation. The name is not just made fearful by its unexpected return in front of our modern consciousness, it was already to the ancients the very "avatar of horror."[101] There is no better example of their awe than the attitude Socrates displays at the beginning of the *Philebus*: even the man who lost his life for introducing "other new demonic beings (*daimonia*)" to the Athenian youth[102] starts his discussion of pleasure, which his adversary proposes to identify with Aphrodite, by avowing a more than human fear in front a divine name, and by therefore invoking the goddess by the name she herself has chosen (12B–C),—a respect neither Psyche nor Aura, as we have seen, are willing to pay, but at their own risk. Thus, as he quotes with enthusiastic approval the transliteration *nemesisch* his friend and correspondent Zelter had used in a previous letter, Goethe is careful to place the hapax under a motto that identifies the goddess with grammar: "grammar takes frightful vengeance at those who despise her."[103] We can then understand how Goethe's friendship with Herder was first shaken due to Herder's unconscionable jokes with his name,—especially the unsavory paronomastic sequence Goethe-*Götter-Gothen-Koth*[104]—a habit that provoked Goethe's bitter retort years later in his autobiography:

> a man's name is not like a mantle, which merely hangs about him, and which, perchance, may be safely twitched and pulled, but is a perfectly fitting garment, which has grown over and over him like his very skin, at which one cannot scratch and scrape without wounding the man himself.[105]

Wilamowitz writes that the word *daimōn* itself deserves to be called demonic because of the number of meanings it has taken over time.[106] I would rather say that a word is demonic insofar as it defies translation, as its transliterated, sphynx-like figure keeps presenting us with the enigma of its meaning. Eric Dodds writes, apropos of the adjective *daimonios*, that already in the *Iliad* its primitive sense has so far faded that Zeus can apply it to Hera. "A verbal coinage so defaced has clearly been in circulation for a long time," Dodds comments[107]: yet the value of a coin is not necessarily its face value.[108]

Freud's essay on *The "Uncanny"* starts with an inquiry in the way other languages express the feeling German refers to as *unheimlich*.[109] But the conclusion he reaches from the perusal of a few bilingual dictionaries is disappointing: "the dictionaries that we consult tell us nothing new," as if any other outcome could be expected to such a search: bilingual dictionaries, of course, do not provide him with a definition of *unheimlich*, but only with translations, they do not offer him any key to the enigma, but only other words no less opaque than the one he is trying to understand. From his foray in foreign territory, however, though no booty, Freud brings back an important insight, as he tries to pinpoint the reason of its failure: the dictionaries tell us nothing new "perhaps only because we ourselves speak a language that is foreign (*vielleicht nur darum nicht, weil wir selbst Fremdsprachige sind*)."[110]

I suggest that we take Freud's awkward statement at its face value here: we ourselves are foreign-speakers, we speak a language that is foreign, namely, to ourselves. From such an insight it would follow that we are no more at ease in our own language than in any other, that any language is equally foreign to us. But Freud is unwilling to go as far, and is quite pleased, after the excursion in uncharted territory, to return on the familiar ground of the German language, having gained "the impression that many languages are without a word for this particular shade (*Nuance*) of what is frightening."[111] The word *unheimlich* is indeed a word that defies translation, though it has not been loaned by any other language but rather replaced by quite uncanny doubles.[112] Freud's translators have thus retroactively proved his hunch that the word is untranslatable. On the other hand, he can report that the search in the German dictionary has brought up something new, namely, Schelling's definition, which "throws a quite new light on the concept of the *Unheimlich*, for which we were certainly not prepared,"[113]—a statement, this time, not to be taken at face value, coming from the initiator of a movement "concerned with laying bare," as Freud puts it in truly "nemesian" terms, "hidden forces."[114]

Freud's own translation of Nemesis is the compulsion to repeat. Rather than Thanatos, however, a name Freud never endorsed, I suggest that Eros's true *Anterōs* in the pantheon (or pandemonium) of psychoanalysis is Ernst Jones's *aphanisis*. Jones remembers the uncanny effect provoked by his introduction of the term in the vocabulary of the new discipline: "Some colleagues," he observes,

> have expressed surprise that just I, who have always insisted on the concrete nature of the unconscious, notably in connection with symbolism, should now describe part of its content by such an abstract Greek term.[115]

Strangely enough, however, Jones does not feel compelled to justify his choice.[116] It seems almost as if, to his mind, only such a formidable term could convey the terrifying fear it is meant to evoke. Jones introduced the term "aphanisis" to refer to a global threat against sexual enjoyment (as opposed to the partial one represented by castration), or the total extinction of sexual desire.[117] Yet the word is no neologism. Jones might have found it used by Plato in the *Sophist* to refer to "the utterly final obliteration (*aphanisis*) of all discourse,"[118] but more probably he recalled it from the Gospels. The hypocrites (*hypokritai*) are condemned by Matth. 6,16 for "disfiguring their faces (*aphanizein ta prosōpa autōn*), that they may appear" different from what they are:[119] in order to disguise themselves, they must first, as it were, blank their countenance.

Even more strangely, maybe, Lacan, who retrieved the term from Jones, proposed to replace it with *fading*, as if in order to neutralize its uncanniness by making it more concrete, a choice that certainly makes things less disquieting to the English speaker.[120] On the other hand, Lacan's intent is to radicalize Jones' notion by reinterpreting it as a criticism of idealism, as it points toward a fault necessarily present in the subject that cannot be dialectically mediated: "when the subject appears somewhere as sense, somewhere else it manifests itself as *fading*, as disappearance;" hence "there is no subject, without, somewhere, *aphanisis* of the subject."[121] Lacan qualifies such a movement of disappearance as "lethal,"—a word that undoubtedly is meant to evoke both the verb *lanthanō*, to hide, and the noun *lēthē*, oblivion, to which it is etymologically related.

Plutarch had exploited the same possibility in his native Greek when he outlined in apocalyptic terms a possible pagan inferno, far more frightening in its sketchiness than any Dante has imagined:

> there is in truth but one penalty for those who have lived ill: obscurity, oblivion, and utter effacement (*adoxia kai agnoia kai pantelōs aphanismos*), which carries them off from Lethê to the joyless river and plunges them into a bottomless and yawning ocean, an ocean that sucks into one abyss all failure to serve or to take action and all that is inglorious and unknown.[122]

The punishment is thus compounded: *aphanismos* is the reward for *aphanisis*, effacement from memory awaits all those who choose to live in hiding, to withdraw from life, following the disengagement advocated by Epicurean teaching. The target of Plutarch's polemic, Epicureanism is, of course, a philosophy of life, but the shameful unconsciously pursue the same aim: to disappear.

Recent clinical literature on shame has, too, recovered Jones's notion and interpreted shame as a defense-mechanism against aphanisis. Shame aims at preventing the consummation of aphanisis by imitating the death of the instincts that aphanisis threatens. What ensues is a "symbolic death," directed at isolating the person from the danger situation. The aim of disappearance may be achieved in different ways:

> most simply, in the form of hiding; most radically, in the form of dissolution (suicide); most mythically, in the form of a change into another shape, an animal or a stone; most archaically in the form of freezing into complete paralysis and stupor; most frequently, in the form of forgetting parts of one's life and one's self; and at its most differentiated, in the form of changing one's character.[123]

Changes of character are thus also a guile of shame, as it were, in order to prevent more catastrophic losses. The highest price one has to pay, however, in order to be able to tolerate the dread of aphanisis, is the disappearance as object of love. Shame is "the ever-deepening conviction of one's *unlovability*."[124] When Rilke renames the prodigal son as "the one who did not want to be loved,"[125] he is truly blaming the shame that hinders his return to the very last: "the shame of having a face."[126]

The conviction of unlovability, however, is not a merely reflexive state of mind, for it denies the very existence of the other. The disappearance as object of love entails the other's disappearance. It is, even more distructively, a threat to Eros itself, and as such cannot be tolerated by Nemesis. Nemesis is invoked by one of the scorned lovers of Narcissus in Ovid's *Metamorphosis*, and "the goddess heard his righteous prayer" (III.406). Proffered in a similar context, Dido's invocation of "a divine power" (*numen*) that has care of the lover not mutually loved (*Aen.* IV.520–521), was interpreted by many commentators, according to Servius, as referring to Nemesis.[127] Thus, when Francesca avows to Dante that "love, which absolves none who are loved from loving (*amor ch'a nullo amato amar perdona*),"[128] still holds her captive, she is still speaking in the name of Nemesis [fig. 24].

Speaking from the point of view of the lover, on the other hand, exposed to the "painful ordeal" of the fading of the loved one, Roland Barthes can only diagnose that "the other seems to lose all desire,"[129]—besides the very desire to disappear. Such a desire manifests itself at first as aphonia. A wonderful instance of the *Tonognomie*, the physiognomy of the voice[130] a contemporary of Lavater, Friedrich Arnold Klockenbring,[131] wished as a necessary supplement to physiognomy, is the observation that the fading of the face is foretold by a fading of the voice. In the diagnosis of the psychiatrist, the lowering of the tone of voice is a spy of the unconscious will to disappear: in such a case "the amount of exhaled air" is "bigger than the sound produced," hence "the sound of horror is always breathy."[132]

The "human form divine" might not resemble, after all, "the face of immortality unveil'd"[133]—such was, perhaps, the "complete knowledge" the dying Kurtz was trying to convey with his final words, as "he cried in a whisper at some image, at some vision," "a cry that was no more than a breath."[134] Yet, like the narrator of Conrad's story, we still have the power of changing a cry to a name.

Nemesis is the power that counteracts Aphanisis. Nothing can remain latent *vis-à-vis* Nemesis, everything must become true in the Greek sense of the unconcealedness (*alētheia*): "exhibitionists," of course, may not "escape the notice of nemesis" (*ouk elathe tēn apasin enantioumenēn tois uperēfanois nemesin*)," but "not even little things elude" her (*oude ta mikra lēthei [. . .] Nemesin*)."[135]

The Epicurean motto: *lathe biōsas*, "live unknown" (Xylander's translation is *latenter vivendum*), seems thus meant especially to challenge the power of the goddess, as already, by way of a paradox, the sophist Gorgias had hoped to "avoid the divine nemesis" while trying to "escape human envy."[136]

Ἀντέρως Amor uirtutis alium
Cupidinem superans.   LXXII.

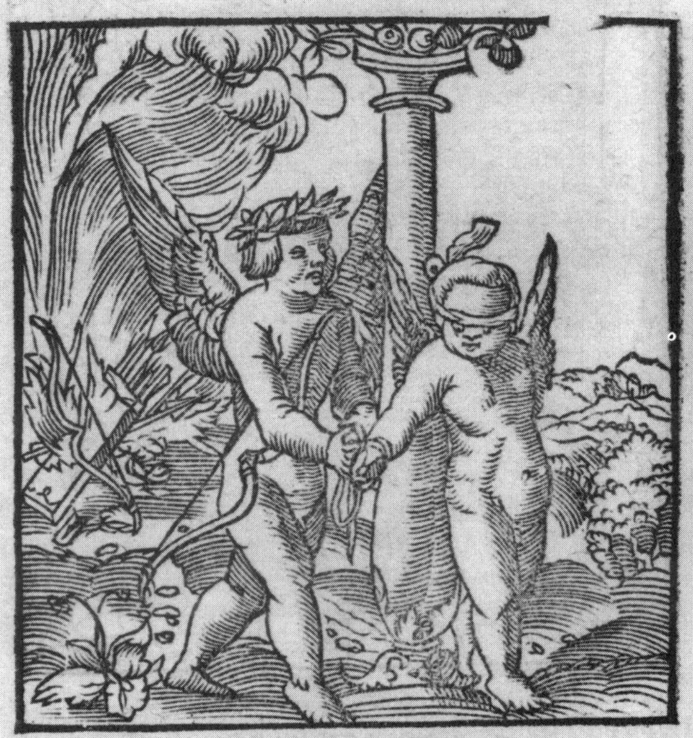

Aligerum aligeroq́; inimicum pinxit Amori,
   Arcu aram, atq́; ignes igne domans Nemesis:
Vt quæ alijs fecit patiatur, at hic puer olim
   Intrepidus gestans tela, miser lachrymat.
Ter spuit inq́; sinus imos (res mira) crematur
   Igne ignis, furias odit Amoris Amor.

Fig. 24. Andrea Alciati, *Emblematum libellus,* Paris 1542. McCormick Library of Special Collections, Northwestern University Library.

We can measure the enduring appeal of the motto, well beyond the fate of ancient Epicureanism, if we consider that Ovid's Latin paraphrasis: "bene qui latuit, bene vixit" (*Trist.* iii. 4, 26), was the epitaph chosen by Descartes for his gravestone,[137] by Nietzsche for his philosophy.[138] The Plutarchean treatise, which questions its wisdom, may be read as a vindication of the goddess. Why should we heed the Epicurean prescription, Plutarch wonders: "is life a shameful thing (*aischron*), that none of us should know about it?"[139] Plutarch argues that genesis does not mean

> to pass into being (*eis ousian odos hē genesis*), as some say, but to pass from being to being known (*ousias eis gnōsin*); for generation does not create the thing generated but reveals it.[140]

Genesis is thus, so to speak, the work of Nemesis, who is constantly engaged in a fight against the opposing drive, which would make of us *déserteurs du monde*.[141] Nemesis counters the fading of the world, as well as our own fading within the world. On the other hand, if Nemesis were to fully overcome her antagonist, the world would then turn into pure display, a mere spectacle for gods and humans alike. Nemesis cannot celebrate her own triumph without bringing upon herself her own wrath. Shamelessness is shame's nemesis.

In Freudian terms, shamelessness "can be understood primarily as a reaction-formation against shame. In the typical 'return of the repressed,' shame merely appears displaced."[142] According to Freud, shame originates in the latency period, as a reaction-formation against sexual impulses, in the same way in which bashfulness is a reaction-formation countering exhibitionistic tendencies.[143] Already in Hesiod the simultaneous departure of Nemesis and Aidos anticipates this dialectic: as they desert the world, they are replaced by "an evil shame" (literally, a not-good shame: *aidōs ouk agathē*),"[144] the "false" or "vitious" shame of the Renaissance translators. Plutarch calls such a time "the extremity of evil" in a fragment of his *Moralia* commenting upon Hesiod's verses. As he often does, Plutarch makes of Plato a spokesperson for his own views: "when asked what conceivable progress his contemporaries had made, [Plato] returned a good answer: 'To have no shame in being wicked.' "[145] Ironically enough, we still measure progress in terms of our attitude to shame: whether we regard human history as a history of growing control over the body, or rather as a history of progressive liberation from the burden of shame,[146] both views ultimately rest upon the humanistic intuition that "verecundia" is the central phenomenon of our life.[147] Not shame, namely, but *verecundia*, the desire to appear as we truly are, yet another pseudonym for Nemesis, but also for Aphanisis, the desire to disappear as what we are not. The hope that leads us to the study of physiognomy, in the formula of a seventeenth-century commentator of Della Porta, Francesco Stelluti, is that the *sfacciati* may become *vergognosi*: that the faceless, i.e., shameless, may become shamefaced. Only by losing (our false) face, we may save (our true) face.[148] Yet shame cannot lend us our true face. Artaud's untranslatable paradox: "le visage humaine n'a pas encore trouvé sa face"—the English translators simply obliterate the difference between *face* and *visage*: "the human face has not yet found its face"[149]—reminds us that, as long as we are ashamed of our face, we will not be able to recognize ourselves.

Carl Friedrich Pockels devotes a chapter in the first volume of his *Essay of a Characteristic of the Female Sex* to the "female discountenance (*Schamhaftigkeit*), the guardian goddess of female virtue,"—a characterization in which are unmistakable the overtones of the edifying language we found used by his contemporary Herder. Pockel distinguishes, by way of a reference to Spinoza, "discountenance (*Schamhaftigkeit*)" from "shame (*Scham*)," the latter being a passion, the former belonging to the realm of virtue.[150] According to Spinoza, a pain may be called good in so far as it indicates that the wounded part is not yet putrified; to the same extent, the man who is discountenanced is at least better than the impudent. Within these limits, although it is not a virtue, as Aristotle and Plutarch had already argued, discountenance is to be considered relatively good.[151]

Like a pain, a word may be said good insofar as it reminds us that we are not yet "mortified."[152] Not "beautiful," like *Schamhaftigkeit*, which so well fits Hans Castorp's character,[153] the word "shame" is the heart of darkness of *The Secret Agent*. More concisely than Kurz's departing breath, Stevie's word, its "inwardness," contains "all his sense of indignation," captures all the horror of the world. The word "shame" becomes itself the record of all shames, a word of which, too, we must be ashamed:

> that little word contained all his sense of indignation and horror at one sort of wretchedness having to feed upon the anguish of the other—at the poor cabman beating the poor horse in the name, as it were, of his poor kids at home.[154]

Stevie shares Nemesis's horror of metonymy, as he realizes that the beating occurs "in the name, as it were," of the poor kids: but only *as it were*.

What survives is the name of shame. Joseph K.'s final thought at the end of *The Trial* goes to shame: "It was as if shame should survive him (*Es war, als sollte die Scham ihm überleben.*)" If shame happen to survive us, let her name at least "bear an ever-during blame;"[155] but let us not abandon hope that we survive her, be it the "shameless hope" by which, Pindar avows, our mortal "limbs are fettered."[156] For only hope can lend us our true, shameless face.

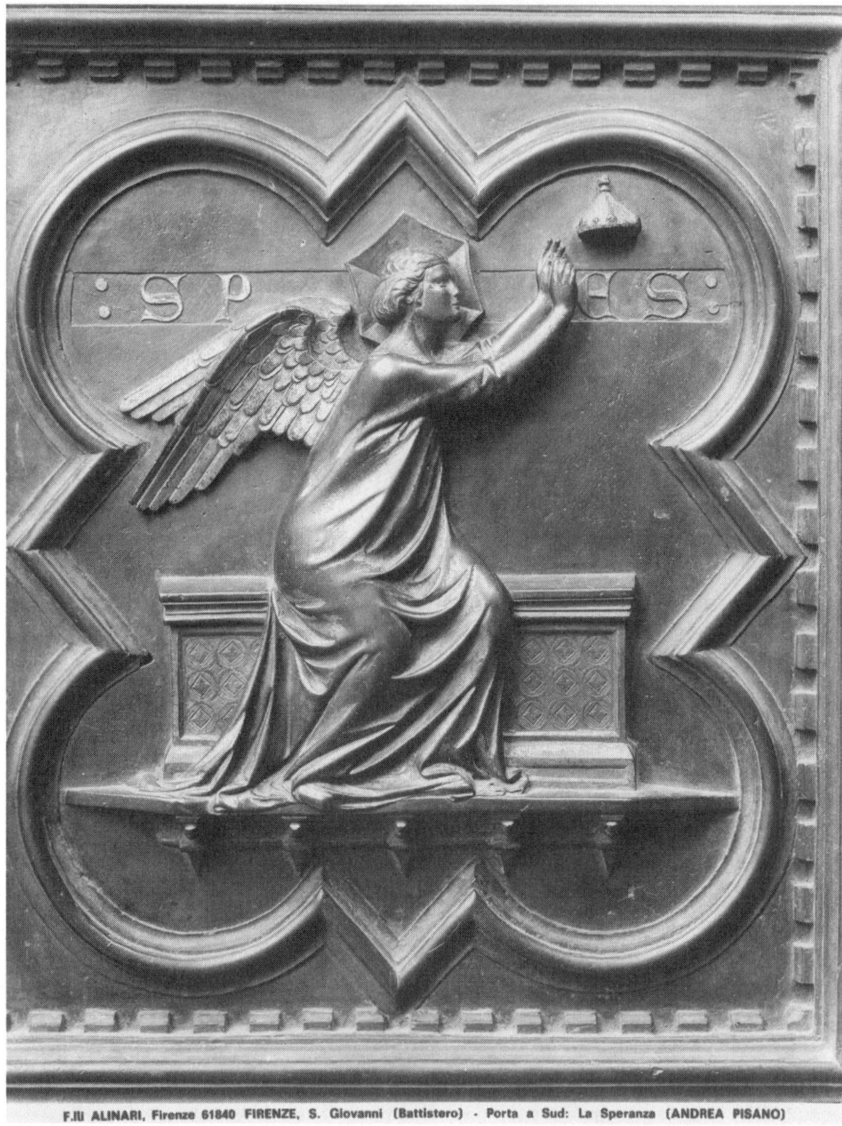

F.lli ALINARI, Firenze 61840 FIRENZE, S. Giovanni (Battistero) - Porta a Sud: La Speranza (ANDREA PISANO)

Fig. 25. Andrea Pisano, *Spes,* Firenze, Baptistery doors. Photo: Archivi Alinari, Firenze.

Helen's appearance on the walls of Troy without the accompaniment of *nemesis* has been interpreted as marking an event of epochal proportions: the redemption of beauty from guilt.[157] Such a redemption is supposedly the work of love: in her second appearance in the poem, Helen is invited by Aphrodite to join Paris, and when she refuses, because such an act would be reprehensible, the goddess of love threatens to abandon her. Helen then follows her *daimōn*.[158] According to such an interpretation, the shield of love protects beauty from *nemesis* only at the price of accepting love as fate. The true redemption of beauty from guilt, however, does not pass through the acceptance of the fatality of love,—hence the submission to its demonic nature—but rather through the hope of the chance of love, *in spite of* its demonic nature. Beauty is a promise of happiness, be it an unattainable one: that is why Elpore, the daughter of Pandora and Epimetheus, Goethe's own personification of hope, is not shy "to promise the impossible."[159] We can recognize her in the *Spes* of Andrea Pisano on the door of the Baptistery in Florence [fig. 25]: "Sitting, she helplessly extends her arm toward a fruit that remains beyond her reach. And yet she is winged. Nothing is more true."[160]

Notes

Introduction

1. Plato, *Charmides* 154B–154D; trans. W. R. M. Lamb in the Loeb Classical Library edition of Plato's works, 12 vols. (Cambridge, Mass.: Harvard UP 1927) 12: 13. I quote Plato, as well as all the other classical authors, from the Loeb Classical Library, unless otherwise indicated.

2. Up to Jacques Derrida's confession that his vision is also impaired by such an ailment: "I am suffering, more and more, from prosopagnosia, a diabolical impulsion to find resemblances in faces, to recognise, no longer to recognise." *La carte postale* (Paris: Flammarion 1980) 203; *Post Card: from Socrates to Freud and beyond*; trans. Alan Bass (Chicago: U of Chicago P 1987) 188.

3. "Die griechischen Plastiker kümmerten sich aber nicht viel um den Kopf, es kam ihnen auf den Körper an, das war vielleicht gerade das Humanistische." *Der Zauberberg*, 2 vols. (1924; Frankfurt/M.: Fischer 1975) 1: 276.

4. In his widely influential *Italian Painters of the Renaissance*, "The Central Italian Painters," first published in 1897 (London: Phaidon 1952) 110–111.

5. Sir Joshua Reynolds, *Discourses on Art*, ed. Robert R. Wark (New Haven: Yale UP 1975) 181.

6. Schelling writes in his *Philosophy of Mythology* that the human figure is "the most accomplished (*la plus finie*)," and that it is thereby chosen as the sign of the apotheosis of Greek divinities (*Philosophie der Mythologie*, in *Sämmtliche Werke* [Stuttgart 1857] II.2: 654, 651).

7. Statius, *Theb.* vi.573; trans. Sir Harry Beaumont (pseud. Joseph Spence), *Crito: or, a Dialogue on Beauty* (London 1752) 14–15. J. H. Mozley's translation for the Loeb Classical Library edition reads: "his face was lost in his body's beauty," 2 vols. (Cambridge, Mass.: Harvard UP 1928) 2: 103.

8. Cf. Adolf Trendelenburg, "Zur Geschichte des Wortes Person," *Kant-Studien* 13 (1908) 15.

9. "Ausspähungskunst des Innern im Menschen" (Immanuel Kant, *Anthropologie in pragmatischer Hinsicht*, *Gesammelte Schriften* [Berlin: Reimer 1907] 285).

10. "Physiognomical QUIXOTISM est MORBUS INSANABILIS!" (*The Physiognomist. A Novel*, by the Author of "The Bachelor and the Married Man." 2 vols. [New York 1820] 2: 271).

11. Of course, an additional, unavoidable ambiguity is due to the circumstance that in the English language "physiognomy" has come to name both the subject of study and the study itself.

12. The merits of books such as Jean-Jacques Courtine's and Claudine Harouche's *Histoire du Visage: Exprimer et taire ses émotions XVIe-début XIXe siècle* (Paris: Rivages 1988) are evident, yet by remaining at the level of a mere archaeology of the body, they beg the question I regard as crucial: which relationship has a face to its own idea, to its *visagéité* (to use the untranslatable term coined by Gilles Deleuze and Felix Guattari: cf. their *Mille plateaux* [Paris: Minuit 1980] 205–234)? This might well have been the original question of Platonism.

13. *The Spectator*, 5 vols. (Oxford: Oxford UP 1964) 1: 366 (June 8, 1711, No. 86). Addison quotes the anecdote from a Latin *florilegium* in support of the claim of physiognomy that "we may be better known by our Looks than by our Words, and that a Man's Speech is much more easily disguised than his Countenance."

14. Cf. Gerhard Neumann, "'Rede, damit ich dich sehe'. Das neuzeitliche Bild und der physiognomische Blick," *Das neuzeitliche Ich in der Literatur des 18. und 20. Jahrhunderts*, ed. Ulrich Fülleborn and Manfred Engel (Munich: Fink 1988) 71–108.

15. Cf. "Wortkunst und Sprachwissenschaft," in his *Stilstudien*, 2 vols. (München: Hueber 1961) 2: 520; and Seneca, *Ad Lucilium Epistulae Morales* cxv.2; trans. Richard M. Gummere, 3 vols. (Cambridge, Mass.: Harvard UP 1925) 3: 320: "style is the garb of thought." In the case of Spitzer, of course, the likelihood that the mispelling be intentional is much higher than in Reynolds's. Since the other motto he selects, *individuum non est ineffabile*, is undoubtedly one of his "parodistic maxims," as Gianfranco Contini defined them, it is not to be excluded that Spitzer intentionally adopted the alternative reading, which better suited his argument. An entry in Ben Jonson's *Timber, or Discoveries* shrewdly brings together Seneca's and Plato's *topoi*: "*Oratio imago animi. Language* most shewes a man: speake that I may see thee." *Ben Jonson*, ed. C. H. Herford Percy and Evelyn Simpson, 11 vols. (Oxford: Clarendon 1947) 8: 625.

16. "Das Wesen des Menschen ist im Gesicht offenbar und in der Sprache verborgen; er kann, weil alles Offenbarwerden für Menschen Schein ist, nur in der Sprache wesenhaft erkannt werden." Werner Kraft, *Franz Kafka: Durchdringung und Geheimnis* (Frankfurt/M.: Suhrkamp 1968) 60.

17. Cf. Giacomo da Lentini's impressive *tour de force* in an untranslatable sonnet built around this pun: "Eo viso, e son diviso da lo viso."

18. Rainer M. Rilke, *Die Aufzeichnungen des Malte Laurids Brigge, Sämtliche Werke*, 6 vols. (Frankfurt/M.: Insel 1966) 6: 711; trans. Stephen Mitchell, *The Notebooks of Malte Laurids Brigge* (New York: Random House 1985) 6, 5.

19. *Paradiso*, Canto XXX, 19–33.

20. Aristotle, *Physics*, 184b10; trans. Philip H. Wicksteed and Francis M. Cornford (Cambridge, Mass.: Harvard UP 1957) 1: 13.

21. Hobbes takes prosopolepsia in the narrow sense, opposite to the one I am advocating, of an "acception of persons," which he denounces as infringing "the fundamental law of nature": "The observance of this law, from the equal distribution to each man, of that which in reason belongeth to him, is called EQUITY, and [. . .] distributive justice: the violation, *acception of persons*, προσωπολημψία." *Leviathan* I.15, in *Works*, ed. William Molesworth (London 1839) 3:142.

22. Henry More, *Enchiridion Ethicum* (2nd. ed., Amsterdam 1679) 66–67 (trans. Edward Southwell, Henry More, *Enchiridion Ethicum: The English Translation of 1690* [New York: The Facsimile Text Society 1930] 90–91). The English translation by Southwell renders More's Latin transliteration (*prosopolepsia*, which he spells like Hobbes, whereas the Greek reads προσωπολημψία, from λαμβάνειν πρόσωπον) with "peculiar fancy." The Hebrew *terminus technicus* for "physiognomy" is *hakkarath panim* (cf. Gershom Scholem, "Ein Fragment zur Physiognomik und Chiromantik aus der Tradition der spätantiken jüdischen Esoterik," *Liber Amicorum: Studies in Honour of Professor Dr. C.J. Bleeker* [Leiden: Brill 1969] 175–193).

23. Cf. Seneca, *Ad Lucilium Epistulae Morales* XL.1, trans. Gummere, 1: 265: "quod in conspectu dulcissimum est, [. . .] agnoscere."

24. See especially his *Totalité et infini: Essai sur l'extériorité* (Hague: Nijhoff 1961).

25. Cf. Gregory Nagy, "Sêma and Nóēsis: Some Illustrations," *Arethusa* 16 (1983) 36–37; Sheila Murnaghan, *Disguise and Recognition in the Odyssey* (Princeton: Princeton UP 1987) 68–70.

26. "ὦθεοί· θεὸς γὰρ τὸ γιγνώσκειν φίλους" *Helen*, ed. A. M. Dale (Oxford: Clarendon 1967). The translation I quote, the most faithful, is by James Michie and Colin Leach (Oxford: Oxford UP 1981) 41, who add the emphasis. Other translations, on the other hand, feel the need to weaken Helen's statement: "Gods!—for God moves in recognition of friends"; trans. Arthur S. Way, 4 vols. (Cambridge, Mass.: Harvard UP 1912) 1:513; "O gods . . . A god *is* at work . . . when we recognize someone dear to us"; trans. Robert Emmet Meagher (Amherst: U of Massachussets P 1986) 38; "I salute the gods. For a god is present when we recognize our loved ones"; trans. James Marwood (Oxford: Clarendon 1997) 135.

27. "Wär nicht das Auge sonnehaft,/wie könnten wir das Licht erblicken?" Goethe's translation of Plotinus's *Enneads* I.6.9 in his "Entwurf zu einer Farbenlehre."

28. *Tintenhaft*: Wilhelm Ostwald, *Goethe, Schopenhauer und die Farbenlehre* (1917; Leipzig: Unesma 1931) 8.

29. Cf. Silvia Rizzo, *Il lessico filologico degli umanisti* (Roma: Storia e Letteratura 1973) 173.

30. "Was nie geschrieben wurde, lesen": Benjamin's definition of physiognomy, borrowed from Hofmannsthal, in his essay "On the Mimetic Faculty" (*Gesammelte Schriften*, 15 vols. [Frankfurt/M.: Suhrkamp, 1974–1989] II·1: 213); in his notes for an essay on "The Task of the Critic," on the other hand, he calls reading "the highest traditional physiognomy (*die höchste traditionelle Physio<g>nomik*)" (*Gesammelte Schriften*, VI: 170). Both definitions, I submit, are consistent with my line of argument.

31. The same verb Euripides uses for "recognizing" was also used in reference to reading as such, with the added prefix *ana-*, which may express the effort of the

reader to understand beyond the mere "recognition" of a text (cf. Jesper Svenbro, *Phrasikleia: An Anthropology of Reading in Ancient Greece*; trans. Janet Lloyd [Ithaca: Cornell UP 1988] 170.) Latin *recognoscere* will become, of course, a technical term in Humanist philology (cf. Rizzo, *Il lessico filologico degli umanisti, passim*).

32. "Auflösung des Rätsels, warum ich niemanden erkenne, die Leute verwechsle. Weil ich nicht erkannt sein will; selber verwechselt werden will." "Materialen zu einem Selbstporträt," *Gesammelte Schriften*, VI: 532. In his fragment "On shame" (*Über die Scham*), Benjamin also points out the mimetic use of blushing as a way of hiding: "in jener dunklen Röte, mit der die Scham ihn übergießt, entzieht sie ihn wie unter einem Schleier den Blicken des Menschen. Wer sich schämt der sieht nichts, allein auch er wird nicht gesehen" (*Gesammelte Schriften*, VI: 70).

33. Léon Wurmser, *The Mask of Shame* (Baltimore: Johns Hopkins UP 1981) 232.

34. Otto Fenichel, *The Psychoanalytic Theory of Neurosis* (New York: Norton 1945) 139.

35. "Ein lächelnder Mund *lächelt* nur in einem menschlichen Gesicht." Ludwig Wittgenstein, *Philosophische Untersuchungen*, § 583.

36. "Incipe, parve puer, risu cognoscere matrem:/Matri longa decem tulerunt fastidia menses./ Incipe, parve, puer, cui non risere parentes/Nec Deus hunc mensa, Dea nec dignata cubili est." Virgil, *Eclogue* IV, vv. 60–63; trans. H. Rushton Fairclough, 2 vols. (Cambridge, Mass.: Harvard UP 1956) 1: 33. The interpretation of these verses is controversial, starting with Quintilian's reading *qui*, rather than *cui*, on line 62, which would entirely put the blame on the children: "those who have not smiled to their parents."

37. "così malvolentieri si parte dal corpo, e ben credo che'l suo pianto e dolore non sia sanza cagione" (Leonardo da Vinci, *Scritti*, Carlo Vecce, ed. [Milano: Mursia 1992] 212). Quoted according to Paul Valéry, "Note and Digression [1919]," *The Collected Works of Paul Valéry*, 13 vols. (Princeton: Princeton UP 1972) 8: 82; Valéry's italics. Cf. also what Leonardo writes in one of his notes toward a treatise on painting: "our soul [. . .] composes the form of the body in which it lives according to its will" ("l'anima nostra [. . .] compose la forma del corpo dov'essa abita, secondo il suo volere," Carlo Pedretti, *Leonardo da Vinci On Painting: A Lost Book (Libro A) Reassembled from the Codex Vaticanus Urbinas 1270 and from the Codex Leicester* [Berkeley: U of California P 1964] 35. I use the English translation by Martin Kemp and Margaret Walker, *Leonardo on Painting* [New Haven: Yale UP 1989] 120.)

38. Valéry, "Note and Digression [1919]," 8: 82–83.

39. G. W. Leibniz, *Nouveaux Essais sur l'Entendement* XXVII.15, *Philosophische Schriften*, ed. C. J. Gerhardt, 6 vols. (Berlin 1882) 5: 223.

40. Valéry, "Note and Digression [1919]," 8: 83. Cf. Robert Klein's comments on a similar sentence by Leonardo in "L'enfer de Ficin," *La forme et l'intelligible* (Paris: Gallimard 1970) 109–110.

41. Cf. Erich Auerbach, "Figura" [1930], in *Gesammelte Aufsätze zur romanischen Philologie* [Bern: Francke 1967] 55–92.

42. Mann, *Zauberberg*, 2: 416.

43. Next to a "physiognomische Kritik" (see below, note 48), Benjamin lists a "strategische Kritik" among the desiderata of criticism in his fragments of an essay

on "The Task of the Critic" (*Gesammelte Schriften*, VI: 172). Cf. also his definition of the critic as "Stratege im Literaturkampf" (*Einbahnstraße, Gesammelte Schriften*, IV.1: 108).

44. Joseph Conrad, *A Personal Record, Complete Works* (New York: Doubleday 1926) 6: 96.

45. Theodor W. Adorno, "Theses upon Art and Religion Today," first published 1945 in the *Kenyon Review*, now in Theodor W. Adorno, *Noten zur Literatur* (Frankfurt/M.: Suhrkamp 1974) 653.

46. Philo, *On the Change of Names* XXIX.155; trans. F. H. Colson and G. H. Whitaker, *Works*, 10 vols. (Cambridge, Mass.: Harvard UP 1929–1939) 5: 221–223.

47. Philo, *On Dreams* I.164, 5: 383: "Might it not have been expected [. . .] that these and like lessons would cause even those who were blind in their understanding to grow keen-sighted, receiving from the most sacred oracles the gift of eyesight, enabling them to judge of the real nature of things, and not merely rely on the literal sense (ὡς φυσιογνωμονεῖν καὶ μὴ μόνον τοῖς ῥητοῖς ἐφορμεῖν)?" Cf. Jean Pépin, *La tradition de l'allégorie de Philon d'Alexandrie à Dante*, vol. 2, *Études historiques* (Paris: Études augustiniennes 1987) 12.

48. "Physiognomische Kritik" (Benjamin, *Gesammelte Schriften*, VI: 172).

49. The double meaning of the German term "Aufgabe," as pointed out by Paul De Man in his essay on "Die Aufgabe des Übersetzers:" "Conclusions: Walter Benjamin's 'The Task of the Translator,' " *The Resistance to Theory* (Minneapolis: U of Minnesota P 1986) 80.

50. *Ep. XX* (*Patrologia Latina*, ed. J.-P. Migne, 221 vols. [Paris 1844] 22: 379).

51. Although a somehow double-edged proposal, if we consider that *litterator,* the native Latin term for a man of letters, was displaced later by *grammaticus*, a transliteration from the Greek (cf. E. M. Bower, "Some technical terms in Roman education," *Hermes* 89 [1961] 477). "Trans-literation" is itself a new coinage, as it was first used (with the hyphen) by Max Müller in a March 9, 1861, article in the *Saturday Review*, discussing the transferral of Buddhist terms from Sanskrit to Chinese.

52. Rather than of the translators, as Valéry Larbaud famously proposed in *Sous l'invocation de Saint Jérome* (Paris: Gallimard 1946).

53. *Rhet.* 1404b2–3; trans. John Henry Freese, *The "Art" of Rhetoric* (Cambridge, Mass.: Harvard UP 1926) 351. Cf. also 1410b12: "Easy learning is naturally pleasant to all, and words mean something, so that all words which make us learn something are most pleasant. Now we do not know the meaning of strange words (*glōttai*), and proper terms we know already" (395–397).

54. Cf. Richard Janko, *Aristotle on Comedy: Towards a reconstruction of Poetics II* (Berkeley: U of California P 1984) 39, 224–225.

55. Cf. *Poetics* 1458a21, 1459a2.

56. O. J. Tuulio, *Ibn Quzman* (Helsinki: Societas Orientalis Finnica 1941) XVIII.

57. G. Scholem, "Reflections on Our Language (*Bekenntnis über unsere Sprache*)," in *Franz Rosenzweig. Zum 25. Dezember 1926 Glueckwuensche zum 40. Geburtstag. Congratulations to Franz Rosenzweig on his 40th Birthday 25 December 1926*, published on the centenary of Franz Rosenzweig's birthday by the Leo Baeck Institute, New York, trans. Martin Goldner (New York 1987) 48.

58. In a letter dated March 10, 1921, to Gershom Scholem, in Franz Rosenzweig, *Briefe und Tagebücher* (Hague: Nijhoff 1979) 1.2: 700. Here Rosenzweig is still under the influence of his friend Eugen Rosenstock, who had argued for a *programmatische Übersetzbarkeit* of foreign words into German (although he had granted to himself an exception, with the word *verecundia*; see below, chapt. 5). On the exchange between Rosenzweig and Scholem on translation cf. Dafna Mach, "Franz Rosenzweig als Übersetzer jüdischer Texte. Seine Auseinandersetzung mit Gershom Scholem," in *Der Philosoph Franz Rosenzweig*, 2 vols. (Freiburg: Alber 1988) 1: 251–271.

59. Ibidem.

60. Freud, *Standard Edition*; trans. James Strachey, 24 vols. (London: The Hogarth Press 1953–1974) VI: 8.

61. As the Latin antiquarian Varro would say, cf. his *De lingua Latina* VII.2 and 45.

62. E. Jones, "The Psychopathology of Everyday Life," *Papers in Psycho-Analysis*, 5th ed. (Boston: Beacon 1961) 43.

63. See below, chapt. 1, p. 26.

64. Cf. Hugo Schuchardt, "The Lingua Franca" (1909), *Pidgin and Creole Languages: Selected Essays*, ed. Glenn G. Gilbert (London: Cambridge UP 1980) 65–88.

65. Proust, *Contre Sainte-Beuve* (Paris: Gallimard 1971) 305: "les beaux livres sont écrits dans une sorte de langue étrangère."

66. Cf. Kafka's speech on the Yiddish language delivered as an introduction to the performance of the actor Jizchak Löwy in the Jewish City Hall in Prague on February 18, 1912: "Er besteht nur aus Fremdwörtern. Diese ruhen aber nicht in ihm, sondern behalten die Eile und Lebhaftigkeit, mit der sie genommen wurden." "Einleitungsvortrag über Jargon" in Kafka's *Kritische Ausgabe, Nachgelassene Schriften und Fragmente I*, ed. Malcolm Pasley (Frankfurt/M.: Fischer 1993) 189.

67. "Die Namen haben ihr Leben." Scholem, *Bekenntnis über unsere Sprache*, 47.

68. Cf. the supremely ambiguous conclusion of his speech, assuring the audience that by awakening the memory of Yiddish "we do not want to punish you." "Einleitungsvortrag," 193.

69. Rosenzweig, *Briefe und Tagebücher*, 1.2: 700.

70. Cf. Benjamin's discussion of the Platonic idea, inspired by Güntert and Usener, in the "Epistemo-Critical Prologue" to *The Origin of German Tragic Drama*; trans. John Osborne (London: NLB 1977) 36–37, and his conclusion that "philosophy is [. . .] a struggle for the representation of a limited number of words which always remain the same—a struggle for the representation of ideas" (37).

71. Cf. Harold Fisch, "The Hermeneutic Quest in *Robinson Crusoe*," in *Midrash and Literature*, ed. Geoffrey H. Hartman and Sanford Budick (New Haven: Yale UP 1986) 213–235: 232.

72. It is correct to specify, as Diego Lanza does in his essay "Quelques remarques sur le travail linguistique du médecin" (*Formes de pensée dans la collection hippocratique; Actes du IV colloque international hippocratique (Lausanne, 21–26 septembre 1981)* [Genève: Droz 1983]), that the physicians and the *physiologoi* never spell out "le projet d'une nomenclature fermée, telle que nous la trouvons par example

dans le premier livre de l'*Histoire des animaux* aristotélicienne" (184). On the other hand, they certainly lay the ground for the formulation of the "programme de définition aristotélicien" (ibid.)

73. As a historian of Arabic physiognomy has pointed out (Youssef Mourad, *La physiognomonie arabe et le Kitâb al-Firâsa de Fakhr al-Dîn al-Râzî* [Paris: Geuthner 1939] 30), at least until the Middle Ages the physician, the physiognomist, the astrologer, and the magician were often just one person—unified in what one might call, with Frazer, the persona of a "public magician" (Sir James G. Frazer, *The Golden Bough: A Study in Magic and Religion*, Part 1: *The Magic Art and the Evolution of Kings*, 2 vols. [3rd. ed; New York: Macmillan 1935] 1: 214–215, 244–247.)

74. From *sōs* e *frēn* (T. G. Tuckey, *Plato's Charmides* [Amsterdam: Hakkert 1968] 5). From a discussion of the various usages of the word in Greek Tuckey concludes that "it is clear that *sōphrosynē* cannot be translated by any one word in English;" on the one hand it means "all that is implied by 'sanity' in the metaphorical sense," on the other "it has a deeper religious significance, to which no one English word can do justice—'humility' expresses some of the meaning, although that word has a connotation possessed by no Greek word" (8–9). Cicero already faced the difficult task in his *Tusculanae*, where he proposes four possible renderings: *temperantia, moderatio, modestia, frugalitas* (III.viii.16–18).

75. On Zalmoxis and shamanism cf. Erwin Rohde, *Psyche: Seelencult und Unsterblichkeitsglaube der Griechen* (Freiburg i. B. 1894) 319–322; Eric R. Dodds, *The Greeks and the Irrational* (1951; Boston: Beacon 1957) 144; Walter Burkert, *Lore and Science in Ancient Pythagoreanism*; trans. Edwin L. Minar, Jr. (Cambridge, Mass.: Harvard UP 1972) 120–165. The shaman "combines the still undifferentiated functions of magician and naturalist, poet and philosopher, preacher, healer, and public counselor" (Dodds, *The Greeks and the Irrational*, 146).

76. 156E; trans. Lamb, 12: 21.

77. The view of Socrates's method as a therapeutics of the soul reaches as far as Marsilio Ficino. In the general introduction to his *Liber de Vita* Marsilio suggests that Socrates, differently from the physician Hippocrates, promises health of soul rather than just of the body: "Sanitatem quidem corporis Hippocrates, animi vero Socrates pollicetur" (Marsilius Ficinus, *Three Books on Life*; ed. and trans. Carol V. Kaske and John R. Clark [Binghamton: Renaissance Society of America 1989] 107). On Plato and medicine cf. at least Werner Jaeger, *Paideia: the Ideals of Greek Culture*, trans. Gilbert Highet, 3 vols. (New York: Oxford UP 1944) 3: 21–27.

78. And in that of the soul, too (cf. David B. Claus, *Toward the Soul: An Inquiry into the Meaning of* ψυχή *before Plato* [New Haven: Yale UP 1981] 170–172).

79. I use these terms in spite of their anachronism in this context: cf. C. Robin, "Recherches historiques sur l'origine et le sens des termes organisme et organisation," *Journal de l'anatomie et de la physiologie* 16 (1880), 1–55; Owsei Tomkin, "Metaphors of Human Biology," *The Double Face of Janus and Other Essays in the History of Medicine* (Baltimore: Johns Hopkins UP 1977) 271–283; Hélène Ioannidi, "Les notions de partie du corps et d'organe," *Formes de pensée dans la collection hippocratique: Actes du IV colloque international hippocratique* (Genève: Droz 1983) 327–330.

80. Kafka, *Nachgelassene Schriften und Fragmente II*, ed. Jost Schillemeit (Fischer, Frankfurt/M. 1992) 62 (trans. Willa and Edwin Muir, *The great wall of China and other pieces* [London: Secker 1933] 293). The Muirs' somehow odd choice of "phiz" for *Fratzengesicht* will not be overly grating in the context of this book.

81. Cf. "Protokollen zu Drogenversuchen," *Gesammelte Schriften*, VI: 588. The theosophical aura is the emanation of a person in the sense in which both breath and smell emanate from a person. It builds the atmosphere surrounding a person. On the influence of the "vulgare Mystik" on Benjamin's concept of the aura, cf. Birgit Recki, *Aura und Autonomie: zur Subiektivität der Kunst bei Walter Benjamin und Theodor W. Adorno* (Würzburg: Königshausen 1988) 43–48.

82. Prosopagnosia is a Freudian term, first introduced in a 1947 essay by Joachim Bodamer, "Die Prosop-Agnosie. (Die Agnosie des Physiognomieerkennens.)," *Archiv für Psychiatrie und Nervenkrankheit* 118 (1948) 6–53, who gives credit to Freud for having put an end to terminological confusion in the field by introducing the concept of "Agnosie" in his essay on aphasia.

83. Quoted in I. A. Richards, *Coleridge on Imagination* (Bloomington: Indiana UP 1960) vii.

84. Cf. the episodes from Ernst Mach's *Analyse der Empfindungen*, and Freud's own recollection, in *The "Uncanny,"* trans. Strachey, *Standard Edition* XVII: 248.

85. *Religio Medici* I. 40. I quote from *The Works of Sir Thomas Browne*, 6 vols., ed. Geoffrey Keynes (London: Faber & Gwyer 1928–1931) 1: 50.

Chapter 1

1. Galen, *Über die medizinischen Namen*, ed. Max Meyerhof and Joseph Schacht, *Abhandlungen der preussischen Akademie der Wissenschaften*, Philosophisch-historische Klasse, n. 3 (Berlin 1931) 10.

2. Kafka's remark, in spite of its avowed unscientificity, but with the first-hand knowledge of the sufferer, is here enlightening: "Looked at with a primitive eye, the real, incontestable truth, a truth marred by no external circumstance (martyrdom, sacrifice of oneself for the sake of another), is only physical pain. Strange that the god of pain was not the chief god of the earliest religions (but first became so in the later ones, perhaps). For each invalid his household god, for the tubercular the god of suffocation." *The Diaries of Franz Kafka 1914–1923*; trans. Martin Greenberg with the cooperation of Hannah Arendt (New York: Schocken 1949) 217–218.

3. Virginia Woolf, "On Being Ill," in *The Moment and Other Essays* (New York: Harcourt Brace Jovanovich 1974) 11.

4. Elaine Scarry writes that "physical pain does not simply resist language but actively destroys it, bringing about an immediate reversion to a state anterior to language, to the sounds and cries a human being makes before language is learned" (*The Body in Pain: The Making and Unmaking of the World* [New York: Oxford UP 1985] 4). On the contrary, it is our language that destroys pain, it is the atrophy of our language, as Virginia Woolf suggests (against Scarry's reading), which has "all grown one way," that inhibits the expression of pain. Woolf's argument suggests that the problem is rooted in the way this language works, and not in the essence of pain; if

at all, then arguably in the essence of language rather than in that of pain. For this failure to understand the historical reason of their untranslatability, Philoctetes's cries must remain sealed to the anonymous English translator Scarry refers to, who renders all of them by the monosyllable "Ah" "followed by variations in punctuations (Ah! Ah!!!!)" (Scarry, *The Body in Pain*, 5), but also unreadable to herself and to all those critics who unachronistically apply to the pain of the ancients the meter of a later stoicism, as Lessing suggests: "I know that we more refined Europeans of a wiser, later age know better how to govern our mouths and our eyes. Courtesy and propriety force us to restrain our cries and tears" (*Laocoön*; trans. Edward Allen McCormick [Baltimore: Johns Hopkins UP 1984] 9). The paradox is not that pain resists language, but that our language distorts pain. Scarry's view of the relationship of language and pain is a document more symptomatic of our own enduring effort at anesthetizing pain than of an intrinsic opacity of pain. In other words, Scarry turns a historical event (the silencing of bodily pain, and its transfiguration in spiritual suffering) into a natural trait of pain, as if this event would not be in itself symptomatic of the process of "misinterpretation of the body" (Nietzsche) that has led to the inexpressibility her book otherwise so eloquently tries to bypass. In a sense, we all ultimately share the destiny of Kierkegaard's poet, "who in his heart harbors a deep anguish, but whose lips are so fashioned that the moans and cries which pass over them are transformed into ravishing music." Soeren Kierkegaard, *Either-Or*; trans. David F. Swenson and Lillian M. Swenson, 2 vols. (Princeton: Princeton UP 1971) 1: 9. Isn't language itself the inescapable Phalaris's bull that amplifies our cries and transforms them into beautiful, soothing melodies? Isn't language as such an attempt to sound *better?*

5. "Thy vertex is Re, thou healthy child, thy temple is Neith, thy eyebrows are the Lord of the East, thy eyes are the lord of Mankind, thy nose is the Nourisher of the Gods, thy ears are the two Royal Serpents, thy elbows are Living Hawks, thy arm is Horus, the other is Seth . . . no limb of thee is without its god, each god protects thy name and all that is of thee . . ." A. Erman, *Zaubersprüche für Mutter und Kind. Aus dem Papyrus 3027 des Berliner Museums*, Berlin 1901; trans. Henry E. Sigerist, *A History of Medicine*, 2 vols., vol. 1: *Primitive and Archaic Medicine* (New York: Oxford UP 1955) 277; cf. Paul Ghalioungui, *The House of Life Per Ankh : Magic and Medical Science in Ancient Egypt* (Amsterdam: Israël 1973]) 38–39.

6. Cf. *De morbo sacro*, esp. chapters I and IV.

7. Origen, *Contra Celsum*; trans. Henry Chadwick (Cambridge: Cambridge UP 1953) 496. Quoting his adversary Celsus, Origen refers to the powers in question as "daemons (*daimones*), or ethereal gods of some sort (*theoi tines aitherioi*)," an identification that cannot be taken for granted, as we will see (cf. chapt. 5, esp. pp. 111 ff.), and betrays the distance of both from an authentic understanding of the doctrine they parody.

8. I transliterate the Greek term, used first by Porphyry, following the example of Singer ("The Scientific Views and Visions of Saint Hildegard (1098–1180)," *Studies in the History and Method of Science*, ed. C. Singer [Oxford: Oxford UP 1917] 38) and Carol Kaske in her "Introduction" to Ficino's *Three Books on Life*, 35.

9. At first only the body of the dead king is identified with the sun-god or the sky-goddess Nut, whose body parts are individual deities. The identification is then

extended from the body of the dead to that of the living king, and from then onward to the body of the nobles and the priests, and finally to that of each Egyptian. Cf. Wilhelm Gundel, *Dekane und Dekansternbilder: Ein Beitrag zur Geschichte der Sternbilder der Kulturvölker, Studien der Bibliothek Warburg*, vol. XIX (Glückstadt: Augustin 1936) 264.

10. Bernard Schultz, *Art and Anatomy in Renaissance Italy* (Ann Arbor: UMI Research Press 1985) 17.

11. Cf. Gundel, *Dekane und Dekansternbilder*, 262.

12. Livy, XX.xxxii; trans. B. O. Forster, slightly modified, 14 vols. (Cambridge, Mass.: Harvard UP 1976) 1: 325. Cf. Jacques Le Goff, "Head or Heart? The Political Use of Body Metaphors in the Middle Ages," *Fragments for a History of the Human Body*, ed. Michael Feher, 3 vols. (New York: Zone 1989) 3:13–26.

13. Cf. Hopkins's line: "Disremembering, dísmémbering áll now," from "Spelt from Sibyl's Leaves," *The Poems of Gerard Manley Hopkins*, ed. W. H. Gardner and N. H. MacKenzie (4th ed., London: Oxford UP 1967). The editors gloss "disremembering" as Irish for "forgetting" (284).

14. Cf. Paul's decisive text, I *Cor.* 12, 14–27: "For the body is not one member, but many. If the foot shall say, Because I am not the hand, I am not of the body; is it therefore not of the body? And if the ear shall say, Because I am not the eye, I am not of the body; is it therefore not of the body? If the whole body *were* an eye, where *were* the hearing? If the whole were hearing, where were the smelling? But now hath God set the members every one of them in the body, as it hath pleased him. And if they were all one member, where were the body? But now are they many members, yet but one body. And the eye cannot say unto the hand, I have no need of thee: nor again the head to the feet, I have no need of you. Nay, much more those members of the body, which seem to be more feeble, are necessary: And those members of the body, which we think to be less honourable, upon these we bestow more abondant honour; and our uncomely parts have more abundant comeliness. For our comely parts have no need: but God hath tempered the body together, having given more abundant honour to that part which lacked: That there should be no schism in the body; but that the members should have the same care one for another. And whether one member suffer, all the members suffer with it; or one member be honoured, all the members rejoice with it. Now ye are the body of Christ, and members in particular."

15. Blaise Pascal, *Pensées* 352, in *Œuvres complètes*, ed. Michel Le Guern, 2 vols. (Paris: Gallimard 2000) 2: 664; trans. Honor Levi, *Pensées and Other Writings* (Oxford: Oxford UP 1995) 90. Sublime, but unfortunately untranslatable, because *esprit du corps* is clearly used by Pascal in a figurative sense.

16. Heinrich Von Staden, *Herophilus: The Art of Medicine in Ancient Alexandria* (Cambridge: Cambridge UP 1989) 89.

17. Fowler's translation ("the nature of the whole man," 1: 549) reduces to the particular case of man what in the original is meant as a statement of methodological importance (cf. A. Diès's discussion of this very problematic passage in *Autour de Platon*, 2 vols. [Paris 1944] 1: 30–45; and the good summary of the debate on its interpretation in Volker Langholf, *Medical Theories in Hippocrates: Early Texts and the 'Epidemics'* [Berlin: De Gruyter 1990] 196–197). On the Hippocratic view of the parts of the body cf. Beate Gundert, "Parts and their Roles in Hippocratic Medicine," *Isis* 83 (1992) 453–465.

18. I prefer to reproduce the Greek wording (*historiē tēs physeōs*) instead of adopting the "old-fashioned," but nonetheless anachronistic translation "natural philosophy," proposed by G. E. R. Lloyd in his *Polarity and Analogy* (Cambridge: Cambridge UP 1966) 12.

19. *Phaedo*; trans. Harold N. Fowler, 1: 339.

20. According to Lloyd, "the sequence 'a capite ad calcem' might be one that any doctor might arrive at independently as soon as he saw the need to introduce some organisation into the account of diseases;" "The Debt of Greek philosophy and Science to the Ancient Near East," *Methods and Problems in Greek Science* (1982; Cambridge UP 1991) 296. But it must have been more than just a convenient classificatory tool, for it clearly enforces a hierarchical organization of the body parts, which assigns to the head the highest rank. Langholf suggests that "perhaps the traditional doctrine about the humours descending from the head and thus causing diseases has prompted this mode of disposition" (*Medical Theories in Hippocrates*, 249).

21. Plato's testimony seems to confirm Bruno Snell's thesis that the early Greeks "did not grasp the body as a unit," but rather as an assemblage of parts; cf. *Die Entdeckung des Geistes: Studien zur Entstehung des europäischen Denkens bei den Griechen* (4th revised ed., Göttingen: Vandenhoeck and Ruprecht 1975) 18; trans. T. G. Rosenmeyer, *The Discovery of the Mind in Greek Philosophy and Literature* (of the first, 1948 German ed.; New York 1953) 5. For more recent literature on the question see Jan Bremmer, *The Early Greek Concept of the Soul* (Princeton: Princeton UP 1983). Snell based his conclusion mainly on philological grounds, namely, on the relative wealth of terms referring to individual limbs and on the absence of a word referring to the body as such in the Homeric poems. In recent years, however, this view and in general Snell's influential account of the "discovery of the mind" in ancient Greece has been taken vehemently to task by Bernard Williams. Williams denounces Snell's implicit assumption of Cartesian parameters as the standard by which to measure the development of ancient theories of the mind and the body. Snell's troubling assumption, in Williams's words, is that "not only in later Greek thought, but truly, a distinction between soul and body describes what we are;" *Shame and Necessity* (Berkeley: U of California P 1993) 25. Williams points out an obvious limit of Snell's approach, which had already been signaled by David Claus, as Williams acknowledges. But the soundness of his criticism does not authorize Williams to conclude that later Greek views of the human body must have been consistent with those of Homeric times, let alone with Williams's own. In support of his criticism, Williams brings the evidence of reading ("every reader of the *Iliad* knows that this cannot be true"), and invokes Priamus's request that Achilles return the body of his son Hector before its dismemberment: "In wanting Hector's body to be whole, Priam wanted Hector to be as he was when he was alive. The wholeness of the corpse, the wholeness that Priam wanted, was not something acquired only in death: it was the wholeness of Hector. Not finding in the Homeric picture of things a certain kind of whole, a unity, where he, on his own assumptions, expects to find one, Snell inferred that what the early Greeks did recognise were merely parts of that whole. In doing this, he overlooked the whole that they, and we, and all human beings have recognised, the living person himself. He overlooked what is in front of everyone's eyes; and in the case of Homer and others of the Greeks, this oversight is quite specially

destructive of their sensibility, which was basically formed by the thought that this thing that will die, which unless it is properly buried will be eaten by dogs and birds, is exactly the thing that one is" (Williams, 24). Williams seems at least no less eager than Snell to find a whole, and his discussion no less vitiated by unacknowledged assumptions—the most questionable of which is the assumption of a continuity in our perception of the world. The assumption of a continuity in our understanding of language that underlies Snell's reasoning is less troubling, albeit also questionable. Williams's argument is founded on the experience of the reader, but even more so on that of a supposedly constant viewer; Snell's, on the experience of the philologist. Snell records the occurrences of a word, Williams sympathizes with a set of mind. I confess that I trust Snell's *akribeia* more than Williams's empathy with the feelings of the Homeric heroes. Again, he charges Snell of not acknowledging "the obvious unity, the one that is in front of his eyes [. . .] The unities needed to have thoughts and experiences are there. They are just the unities that Homer's characters recognised as thinking and feeling: themselves" (Williams, 26). The case of the soul, however, in Williams's judgment, is more complicated than that of the body, and this is due to the fact that "we do indeed have a concept of the body, and we agree that each of us has a body. We do not, *pace* Plato, Descartes, Christianity, and Snell, all agree that we each have a soul. Soul is, in a sense, a more speculative or theoretical conception than body" (Ibidem.) But in *which* sense is the body a less speculative conception than the soul? Maybe just because we did not think as much about it. These generalizations are no less problematic than Snell's, and confirm the validity of Nietzsche's hypothesis, that "the belief in the body is more fundamental than the belief in the soul;" but, hence, even so much harder to dispel. A more convincing answer to the question of personal identity than Williams's ostensive gesture is given by Albin Lesky in a passage Williams quotes without comments in a footnote: "The simplest expression, and one which precedes all abstraction, for the identity of the person preserved through all the phasis of the action is the proper name" (*Der einfachste, aller Abstraktion vorausliegende Ausdruck für die durch alle Phasen der Handlung festgehaltene Identität der Person ist der Eigenname.*) "Göttliche und menschliche Motivation im homerischen Epos," *Sitzungsberichte der Heidelberger Akademie der Wissenschaften, Philosophisch-historische Klasse* 45 (1961) 11. But this quote, which alludes to the preserved unity of the character throughout the unfolding of the epic plot, contradicts Williams's own train of argument: the name is the most immediate guarantee of identity for the Homeric characters, certainly not the integrity of the body. The name is the token of the unity of the person, not the body, in a universe in which names have guarded their magical power, the power to name (cf. P. M. Schuhl, *Essai sur la formation de la pensée greque* [Paris: Vrin 1934] 42; Johannes Lohmann, *Musiké und Logos: Aufsätze zur griechischen Philosophie und Musiktheorie* [Stuttgart: Musikwissenschaftliche Verlags-Gesellschaft 1970] 14). Cf. below, *passim.*

22. J. B. deC. M. Saunders, *The Transition from Ancient Egyptian to Greek Medicine* (Lawrence: University of Kansas Press 1963) 19. Cf. Ludwig Edelstein, "Hippocratic Prognosis" (1931); trans. C. Lilian Temkin, in L. Edelstein, *Ancient Medicine: Selected Papers*, ed. Owsei Temkin and C. Lilian Temkin (Baltimore: Johns Hopkins 1967) 66: "There is in ancient medicine no such theory of a disease *per se*, independent of the affected organ."

23. G. E. R. Lloyd, "Introduction," *Hippocratic Writings* (Harmondsworth: Penguin 1986) 21. This is not, of course, the only source of the vocabulary of pathology; another powerful resource is analogical thinking: "the names of diseases are often taken from things in nature to which they bear a resemblance" (Lloyd, *Polarity and Analogy*, 177).

24. Hermann Usener, *Götternamen: Versuch einer Lehre von der religiösen Begriffsbildung* (Bonn 1896) 317–318. On Usener's theory cf. Ernst Cassirer, *Sprache und Mythos: Ein Beitrag zum Problem der Götternamen, Studien der Bibliothek Warburg* VI (Lepzig: Teubner 1925), esp. pp. 12–20.

25. Reynolds, *Discourses on Art*, 180.

26. Ralph W. Emerson, "Introduction," *Plutarch's Miscellanies and Essays*, ed. William W. Goodwin, 5 vols. (1870; 6th ed., Boston: Little, Brown, and Company 1889) 1: xvi. A reference to Plato's *Phaedon*, 83D: "each pleasure or pain nails it [the soul] as with a nail to the body" (trans. Fowler, 1: 291). The same image returns in *Timaeus* 43A and has been transmitted to the Middle Ages through Calcidius's commentary (cf. C. S. Lewis, *The Discarded Image: An Introduction to Medieval and Renaissance Literature* [1964; Cambridge 1974] 60).

27. Cf. Heraclitus fr. 103 Diels. Kirk's edition lists other references to this principle in the Hippocratic corpus, all to the point that "attacks on disease must be made through the body as a whole" (Heraclitus, *The Cosmic Fragments*, ed. G. S. Kirk [2nd rev. ed.; Cambridge: Cambridge UP 1962] 114).

28. Hippocrates, *De locis in homine* I.1–I.5; trans. Paul Potter, *Places in Man*, in the Loeb Classical Library edition of Hippocrates works, 8 vols. (Cambridge, Mass.: Harvard UP 1923–1995) 8: 19–21.

29. XLIV–XLVI. Cf. Robert Joly's "Notice," *Des Lieux dans l'homme* (Paris: Belles Lettres 1978) tome XIII: 28–31.

30. Hippocrates, *On Breaths* II; trans. W. H. S. Jones, 2: 229.

31. Snell, 16.

32. "some say it is the tomb (*sēma*) of the soul, their notion being that the soul is buried in the present life; and again, because by its means the soul gives any sign which it gives, it is for this reason also properly called sign (*sēma*). But I think it most likely that the Orphic poets gave this name, with the idea that the soul is undergoing punishment for something; they think it has the body as an enclosure to keep it safe, like a prison, and this is, as the name itself denotes, the safe (*sōma*) for the soul, until the penalty is paid, and not even a letter needs to be changed" (trans. Fowler, 4: 63) On the relationship between the two meanings of the word *sēma*, cf. Gregory Nagy, "Sêma and Nóēsis: Some Illustrations," *Arethusa* 16 (1983), 35–55, esp. 45ff.

33. On Baylonian physiognomy cf. Fritz Rudolf Kraus, *Die physiognomischen Omina der Babylonier, Mitteilungen der Vorderasiatisch-Aegyptischen Gesellschaft*, 40/2 (Leipzig: Hinrichs 1935). On the origins of Greek physiognomy, cf. Richard Foerster, "Prolegomena," *Scriptores physiognomonici Graeci et Latini*, 2 vols. (Lipsiae 1893) VII–XV; Johannes Thomann, "Anfänge der Physiognomik zwischen Kyoto und Athen: Sokratische Begriffsbestimmung und aristotelische Methodisierung eines globalen Phänomens," *Die Beredsamkeit des Leibes: Zur Körpersprache in der Kunst*, ed. Ilsebill Barta Fiedl and Christoph Geissmar (Salzburg: Residenz 1992) 209–215.

34. Woolf, "On Being Ill," 11.

35. *Prior Analytics* 70b7ff. Lloyd rightly points out the still hypothetical nature of this endorsement (G. E. R. Lloyd, *Science, Folklore and Ideology: Studies in the Life Sciences in Ancient Greece* [Cambridge: Cambridge UP 1983] 22–23), whereas the anonymous compiler of the pseudo-Aristotelian physiognomy dismisses any doubt: "now if this is true (and it is invariably so), there should be a science of physiognomics" (805a17–18); trans. W. S. Hett, *Physiognomics*, in Aristotle, *Minor Works* (Cambridge, Mass.: Harvard UP 1936) 85.

36. Cf. esp. Elmar Siebenborn, *Die Lehre von der Sprachrichtigkeit und ihren Kriterien: Studien zur antiken normativen Grammatik* (Amsterdam: Grüner 1976) 116–139; Walter Belardi, *Filosofia, grammatica e retorica nel pensiero antico* (Roma: Ateneo 1985) 9–20.

37. Taking "anatomy" in a broader, metaphorical sense, as already Longinus (XXXII.5, apropos of Xenophon and Plato) and then most humanists did; e.g., Melanchthon: "descriptio humani corporis, seu ut vocant ἀνατομή." *Commentarius de anima* (Wittemberg 1540) 32r.

38. Cf. Belardi, *Filosofia, grammatica e retorica*, 12.

39. Cf. for the Semitic languages Harri Holma, *Die Namen der Körperteile im Assyrisch-Babylonischen: Eine lexical-etymologische Studie* (Leipzig 1911) VIII.

40. Fragm. 1 Diels (Hermann Diels-Walther Kranz, *Die Fragmente der Vorsokratiker*, 6th ed., 3 vols. [Berlin: Weidmann 1952] 2: 262–264).

41. Leon Battista Alberti, *On Painting and on Sculpture: The Latin Texts of De Pictura and De Statua*, ed. and trans. Cecil Grayson (London: Phaidon 1972) 53.

42. Cf. Hans-Georg Gadamer, "The Expressive Power of Language: On the Function of Rhetoric for Knowledge"; trans. Richard Heinemann and Bruce Krajewski, *PMLA* 107 (1992) 349–350; originally published as "Die Ausdruckskraft der Sprache: Zur Funktion der Rhetorik für die Erkenntnis," *Lob der Theorie* (Frankfurt/M.: Suhrkamp 1983).

43. Italian has a word to refer to this quality of the body, its commensurability: *corporatura.*

44. Trans. Fowler, 1: 529. According to Theodor Gomperz, *Griechische Denker* (1903) 2: 574; cit. in Elisabeth Rotten, *Goethes Urphänomen und die platonische Idee* (Gießen: Töpelmann 1913) 102. But Creuzer had already drawn this genealogy in his commentary to Plotinus's treatise *On Beauty* (*Plotini Liber de Pulcritudine*, ed. Friedrich Creuzer [Heidelberg 1814] 153.)

45. Trans. Fowler, 1: 529.

46. *Theaetetus* 207A; trans. Fowler, 2: 243.

47. *Theaetetus* 204A, 2: 231.

48. *Sophist* 253A, 2: 399.

49. 262A, 2: 435.

50. 259E, 2: 425–427.

51. 262D, 2: 437.

52. 253A, 2: 399. According to Jaeger, who, however, only refers to *Timaeus* 31B–C, Plato is the first to have used the term *desmos* figuratively (cf. Werner W. Jaeger, *Nemesios von Emesa: Quellenforschungen zum Neuplatonismus und seinen Anfängen bei Poseidonios* [Berlin: Weidmann 1914] 101). Dante calls the vowels "anima e legame d'ogni parole" in *Convivio* IV.vi.

53. *Rhetoric* 1407a20. Cf. Rudolf Pfeiffer, *History of Classical Scolarship: From the Beginnings to the End of the Hellenistic Age* (Oxford: Oxford UP 1968) 76–77; Belardi, *Filosofia, grammatica e retorica*, 131–145; Peter Matthews, "La linguistica greco-latina," *Storia della linguistica*, ed. Giulio C. Lepschy, 2 vols. (Bologna: Mulino 1990) 1: 218–220.

54. "The well-known epic pattern of the description of human beauty which consists in listing the parts of the body from head to foot" is mentioned by Leo Spitzer "as an illustration of the medieval habit of placing any phenomenon within a framework, within a closed whole" (*Essays in Historical Semantics* [New York: Vanni 1948] 269). But this habit is far more ancient than the Middle Ages, and it goes indeed, once again, as far back as the ancient Egyptians: cf. Alfred Hermann, "Beiträge zur Erklärung der ägyptischen Liebesdichtung," *Ägyptologische Studien*, ed. O. Firchow (Berlin: Akademie-Verlag 1955) 118–139, esp. 124–133, who sees the origins of the descriptive poem in the cultual hymns meant as apotheosis of the deceased regent.

55. "Quoniam igitur et signa membrorum et significationes ipsae signorum propemodum expositae atque enumerate sunt tamquam prima elementa [. . .] litterarum, nunc concipiamus atque constituamus species aliquas ex pluribus, sicut ex litteris syllabae constituuntur [. . .] Constituamus uirum fortem." Anonyme Latin, *Traité de physiognomonie*, ed. Jacques André (Belles Lettres 1981) 121. Cf. Antoine du Moulin, *De diversa hominum natura* (Lugduni 1549) 11.

56. Ioannis Padovanii Veronensis, *De Singularum Humani Corporis Partium Significationibus* (Veronae 1589).

57. Venetiis 1609.

58. "The blasonneurs quickly found that the female anatomy is made up of a relatively limited number of component parts, and clearly experienced some difficulty in finding suitable subjects which had not already been treated. Gilles d'Aurigny succinctly describes this problem at the beginning of his *Ongle* (ca. 1545): Il n'y a si gentil esprit/qui n'ai inventé ou escript/Quelque chose à l'honneur du corps:/J'entends des membres des dehors,/Tant que plusieurs qui s'y sont mis,/Pensent que rien n'y soit obmis./Chacun a faict blason honneste,/Depuis le pied jusqu'à la teste." Alison Saunders, *The Sixteenth-Century Blason Poétique* (Bern: Lang 1981) 126. The blason, as the Renaissance form par excellence of the "epic description of beauty," is particularly interesting in this context. The earliest edition of the blasons were printed from 1536 onwards as an appendix to Alberti's *Hecatomphile* (Paris 1536; first Italian ed. 1471 Padova) (Saunders 113), but the fashion probably originated with Olimpo da Sassoferrato's *Gloria d'amore* (1529), a collection of forty-five strambotti devoted to the various features of the lady, under the general title *Comparation de laude a la signora mia incominciando al capo per insino a li piedi* (Saunders 94.) Cf. Roland Barthes's assessment of the blason-genre, which is fully consistent with my argument in this chapter: "Comme genre, le blason exprime la croyance qu'un inventaire *complet* peut reproduire un corps *total*, comme si l'extrême de l'énumération pourait basculer dans une catégorie nouvelle, celle de la totalité" (*S/Z* [Paris: Seuil 1970] 121).

59. Galen, *de plac.Hipp. et Plat.* 5 (*Stoicorum Veterum Fragmenta* III, fr. 472).

60. Quoted in John Murdoch, "From Social into Intellectual Factors: An Aspect of the Unitary Character of Late Medieval Learning," *The Cultural Context of Medieval Learning*, ed. John Murdoch and Edith Sylla (Dordrecht: Reidel 1975) 342.

61. "corpus humanum est corpus mistorum dignissimum omnium, & vinculum universi" (*Conciliator controversiarum, quae inter philosophos et medicos versantur, Petro Abano Patavino,Philosopho ac Medico clarissimo Auctore* [Venetiis 1548] 10r).

62. Cf. Walter Pater, "Pico della Mirandola," *The Renaissance: Studies in Art and Poetry* (Oxford: Oxford UP 1986) 26.

63. *Timaeus* 87C–D; trans. R. G. Bury, 9: 237–239.

64. Cf. Jan Bialostocki, *Dürer and His Critics* (Baden-Baden: Koerner 1986), chapt. II, esp. 23ff.

65. Vicente Carducho lists Gallucci as first among the physiognomists whose study he recommends to the painter in his *Dialogos de la pintura* (1633; Madrid: Turner 1979) 30.

66. "Alberto Durero [. . .] di gran lunga superò tutti quelli, che avanti lui (quantunque siano da Historie, & versi celebratissimi) furono, & ai posteri lasciò se stesso ne i suoi scritti, & disegni idea della vera Pittura, & della Scoltura, come chiaramente si vede si nelle carte, si in questo libro della simmetria de i corpi humani." *Di Alberto Durero pittore, e geometra chiarissimo. Della simmetria dei corpi humani, Libri Quattro. Nuovamente tradotti dalla lingua Latina nella Italiana, da M. Gio. Paolo Gallucci Salodiano. Et accresciuti del quinto libro, nel quale si tratta, con quai modi possano i Pittori, & Scoltori mostrare la diversità della natura de gli huomini, & donne, & con quali le passioni, che sentono per li diversi accidenti, che li occorrono. Hora di nuovo stampati.* (Venezia 1591) f. †2v.

67. F. †1v.

68. Dolce uses the form "Duro" in his "Dialogo della Pittura" (1557), edited and trans. by Mark W. Roskill in his *Dolce's "Aretino" and Venetian Art Theory of the Cinquecento* (New York: New York UP 1968); where he precisely criticizes "Dürer's sense of propriety (*convenevolezza*) [. . .] not only in the case of costumes, but also in the case of faces (*volti*)" (120–121). Cf. Bialostocki, *Dürer and His Critics*, 63.

69. Erwin Panofsky, *The Life and Art of Albrecht Dürer* (4th ed.; Princeton: Princeton UP 1955) 273. Cf. also Peter W. Parshall, "Camerarius on Dürer—Humanist Biography as Art Criticism," *Joachim Camerarius (1500–1574): Beiträge zur Geschichte des Humanismus im Zeitalter der Reformation*, ed. Frank Baron (Munich: Fink, 1978) 11–29, esp. 11–12.

70. "non hauendo egli lasciato particella alcuna de i nostri corpi esteriore però (non considerando altro del huomo il Pittore, & lo Scoltore, che quello, che si vede) che esso non habbia misurata, & col suo divino ingegno spiegata con tanta sottigliezza, che fa stupire chiunque è di quell'arte studioso, e perito," *Della simmetria dei corpi humani*, f. †2v.

71. "cum autor curiosa pene diligentia exquisiverit partium in corpore humano nomina, quo mensurationes certiores essent, quibusdam etiam nova imposuerit, confido fore ut studiosi versionis vel hac in parte difficultatem intelligant, nam reliqua praetereo quae et ipsa non possint facilia videri fuisse, cum in hoc genere quod imitaremur, antiquorum extaret nihil. Conquisivimus autem et nos non mediocri cura neque modico tempore nomina quibus reddere Düreriana possemus, quae quam apposita sint, non nos quidem praefari decuerit, sed legentium erit iudicium." *Alberti Dureri clarissimi pictoris et Geometrae de Symmetria partium in rectis formis humanarum corporum Libri in latinum connessi.* (Nürnberg 1532) f. A4ᵛ. Reprinted in Dürer's *Schriftlicher*

Nachlaß, ed. Hans Rupprich, 4 vols. (Berlin: Deutscher Verein für Kunstwissenschaft 1956f.) 1: 312.

72. F. A5r (not reprinted by Rupprich).

73. *Ioachimi Camerarii Pabeperg. Commentarii utriusque Linguae, in quibus est ΔΙΑΣΚΕΥΗ ΟΝΟΜΑΣΤΙΚΗ ΤΩΝ ΕΝ ΤΩ ΑΝΘΡΩΠΙΝΩ ΣΩΜΑΤΙ ΜΕΡΩΝ, hoc est, Diligens exquisitio nominum, quibus partes corporis humani appellari solent* (Basileae 1551).

74. Cf. Patrizia Landucci Ruffo, "Le fonti della "Medicina" nell'Enciclopedia di Giorgio Valla," *Giorgio Valla tra scienza e sapienza* (Firenze: Olschki 1981) 55–68; Giovanni Pozzi, "Appunti sul "Corollarium" del Barbaro," *Tra latino e volgare: Per Carlo Dionisotti* (Padova: Antenore 1974) 2: 638: "È la volontà di chiamare tutte le cose nel loro autentico nome antico e perciò di esplorare tutto l'antico tesoro lessicale per scoprirvi ogni sfumatura linguistica."

75. K. B. Roberts and J. D. W. Tomlinson, *The Fabric of the Body: European Tradition of Anatomical Illustration* (Oxford: Oxford UP 1992) 49.

76. *Anatomia Philologica, continens Discursus Philologicus De Nobilitate et Praestantia Hominis Contra iniquos conditionis humanae aestimatores Autore Gregorio Queccio.* (Nürnberg 1632). This work is a *centone* in the tradition of the apologetical literature on the dignity of man.

77. Philologically tested terms could now be at last retrieved out of a revised Aristotelian text or from the numerous editions of the Greek physicians that were published in this century. Particularly important in this respect were the new editions of Pollux's *Onomasticon*, first edition 1502 in Venice by Aldus, 1542 Latin translation by Rudolph Gualterus in Basel (cf. DeWitt T. Starnes, *Renaissance Dictionaries English-Latin and Latin-English* [Austin: U of Texas P 1954] 170–171; the first edition was preceded by the publication of Giorgio Valla's *De humani corporis partibus* in 1501, which introduces new anatomical terms from Pollux; cf. Charles Singer, "The Confluence of Humanism, Anatomy and Art," *Fritz Saxl, 1890–1948*, ed. Donald J. Gordon [London: Nelson 1957] 268); Rufus's *De appellationibus partium corporis humani*, Latin translation by Iulius Paulus Crassus in the collection of *Medicae artis principes, post Hippocratem et Galenum* (Basileae 1567); Theophilus's *De Corporis humani fabrica*, translated by Crassus in the collection *Medici Antiqui Graeci* (Basileae 1581); and Meletius's *De Natura structuraque hominis*: *Meletii Philosophi De Natura Structuraque hominis* (Venetiis 1552). This process culminated in the compilation of works such as Bartolomeo Castelli's *Lexicon Medicum, Graecolatinum* (Messanae 1598); first published at the end of the sixteenth century, with the addition of a glossary of Arabic terms with the corrispective Greek and Latin equivalents, this will become an indispensable instrument of consultation for at least the following two centuries. Repeatedly reprinted and revised (1651, 1657, 1688, 1713, 1746).

78. "Graeci nihil relinquerunt sine nomine, gens dives vocum," *Commentarii utriusque Linguae*, c. 125.

79. "in dictionibus quibusdam transferendis, nec mihi ipsi satisfacere potuerim, quod quidem non tam culpa propria, quam Latinae linguae accidit penuria." *ΘΕΟΦΡΑΣΤΟΥ ΚΑΡΑΚΤΗΡΕΣ. Cum interpretatione Latina per Bilibaldum Pirckeymherum, iam recens aedita* (Nürnberg 1527) f. a2r. Rpt. in Dürer, *Schriftlicher Nachlaß*, 1: 119.

80. Cf. Walter Kambartel, *Symmetrie und Schönheit* (München: Fink 1972). Hermann Weyl is inaccurate when he takes the word "symmetry" to be still used in our everyday language in two concurring meanings: one as a synonym of "harmony" (or German "Ebenmass"), the other as bilateral symmetry, "a strictly geometric [. . .] concept" (Hermann Weyl, *Symmetry* [1952; Princeton: Princeton UP 1969] 6). The examples Weyl submits, from Polykleitos to Vitruvius to Dürer, show that the first meaning is hardly a current one. Coleridge still uses the word in the ancients' sense, when he writes: "the symmetry of a body results from the sanity and vigour of the life as the organizing power" (cit. in M. H. Abrams, *The Mirror and the Lamp: Romantic Theory and the Critical Tradition* [Oxford: Oxford UP 1953] 224); whereas Ruskin precisely marks the passage from the ancient to the modern sense of symmetry, as opposed to proportion, when he writes that "symmetry is the *opposition* of *equal* quantities to each other. Proportion the *connection* of *unequal* quantities with each other;" and exemplifies the distinction by pointing out that "in the human face its balance of opposite sides is symmetry, its division upwards, proportion." "It seems strange," Ruskin concludes, "that the two terms could ever have been used as synonymous" (*Modern Painters*, 4 vols. [New York 1883] 2: 72–73).

81. Signally in Boethius, *Musica* I.31; cf. Pomponius Gauricus, *De Sculptura (1504)*, ed. André Chastel and Robert Klein (Genève: Droz 1969) 73; Michael Baxandall, *Patterns of Intention: On the Historical Explanation of Pictures* (New Haven: Yale UP 1985) 112. Gellius, *Noctes Atticae* I.i uses "competentia" (which Calcidius interprets as *analogia*), Pliny the younger "congruentia" and "aequalitas."

82. "quae Graece ἀναλογία, Latine—audendum est enim, quoniam haec primum a nobis novantur—comparatio pro portione dici potest." Cicero, *Timaeus* 13.

83. "Ea autem paritur a proportione, quae graece *analogia* dicitur. Proportio est ratae partis membrorum in omni opere totiusque commodulatio, ex qua ratio efficitur symmetriarum. Namque non potest aedis ulla sine symmetria atque proportione rationem habere compositionis, nisi uti ad hominis bene figurati membrorum habuerit exactam rationem." Trans. Frank Granger, *On Architecture*, 2 vols. (Cambridge, Mass.: Harvard UP 1962) 1: 159. *Symmetria* was for Vitruvius one of the elements of architecture, along with *ordinatio, dispositio, eurythmia, decus,* and *distributio.* He gives a first definition of *symmetria* in I.ii.4: "Symmetry is the appropriate harmony (*conveniens consensus*) arising out of the members of the work itself; the correspondence (*responsus*) of each given part among the separate parts to the form of the design as a whole (*ad universae figurae speciem*)."

84. *Di Lucio Vitruvio Pollione de Architectura Libri Dece traducti de latino in Vulgare* (Como 1521) 48r.

85. "Tradizione classica e volgarizzamenti," *Geografia e storia della letteratura italiana* (Torino: Einaudi 1967) 166. On Cesariano's translation, which Olschki mercilessly labels "das barbarischste Buch des gesamten italienischen Schrifttums" (*Geschichte der neusprachlichen wissenschaftlichen Literatur*, vol. 2: *Bildung und Wissenschaft im Zeitalter der Renaissance in Italien* [Leipzig: Olschki 1922] 203) cf. also Manfredo Tafuri, "Cesare Cesariano e gli studi vitruviani nel Quattrocento," *Scritti rinascimentali di architettura* (Milano: Polifilo 1978) 387–438; and Rudolph Wittkower, *Architectural Principles in the Age of the Humanism* (New York: Norton 1971) 14–15.

86. Leon Battista Alberti, *L'Architettura [De re aedificatoria]*, 2 vols. (Milano: Polifilo 1966) 2: 441; trans. James Leoni, *The Ten Books of Architecture* (1755; New York: Dover 1986) 111.

87. *Trattati di architettura ingegneria e arte militare*, 2 vols. (Milano: Polifilo 1967) 2: 295.

88. Cf. Leonardo's epitaph by Platino Piatto: "Mirator veterum, discipulusque memor,/ Defuit mihi symmetria prisca. Peregi/Quod potui; Veniam da mihi, posteritas." Cit. according to Vernon Lee, "Symmetria Prisca," *Euphorion: Being Studies of the Antique and the Mediaeval in the Renaissance*, 2 vols. (London: Unwin 1884) 1: 169.

89. ". . . e le figure ne' superiori pittori morte fece vive e di vari gesti." Cristoforo Landino, "Proemio al commento dantesco" (1481), *Scritti critici e teorici*, 2 vols., ed. Roberto Cardini (Roma: Bulzoni 1974) 1: 123–124. Alberti also feels the need to stress the Greek origin of the word in the Latin version of *On Painting*: "symmetriam, ut Graeci aiunt" (*On Painting*, 98); whereas he straightforwardly translates it into Italian as "misura" (*Opere*, 96).

90. Leonardo Bruni, *De interpretatione recta* (ca. 1420), cit. in Gianfranco Folena, "'Volgarizzare' e 'tradurre': idea e terminologia della traduzione dal Medio Evo italiano e romanzo all'Umanesimo europeo," in *La traduzione: saggi e studi* (Trieste: Lint 1973) 99.

91. He is credited for producing, among others, the first dependable translation of Aristotle's *Poetics* (1498), on which cf. Bernard Weinberg, *History of Literary Criticism in the Italian Renaissance*, 2 vols. (Chicago: U of Chicago P 1961) 1: 361–366. On Valla's (1447–1500) remarkable personality see the essays collected in *Giorgio Valla tra scienza e sapienza* (Firenze: Olschki 1981); Carlo Dionisotti, *Gli umanisti e il volgare fra quattro e cinquecento* (Firenze: Le Monnier 1968) 44–47; Cesare Vasoli, *La dialettica e la retorica dell'Umanesimo* (Milano: Feltrinelli 1968) 132–144. On his translations in the field of musical theory cf. Claude V. Palisca, *Humanism in Italian Renaissance Musical Thought* (New Haven: Yale UP 1985) 67–87.

92. "Symmetria, quae latine commensurabilitas dici potest, est qua magnitudines aliquae sub mensuram aliquam cadunt, sicut contra Asymmetria quae sub mensuram communem aliquam non cadunt" (Venetiis 1501) f. LLiiiv. On the following folio he then reproduces word for word Vitruvius's definition from the first book.

93. Cf. Lilian Defradas, "Les sources du *De physiognomonia* de Pomponius Gauricus," *Bibliothèque d'Humanisme et Renaissance* 32 (1970) 7–39.

94. Gauricus, *De Sculptura*, 73, 93.

95. Gauricus 97: "Consideranda vero et ipsa inter se partium ἀναλογία, quam alibi proportionem, heic ni fallor proprie commensum dixerimus."

96. 13r.

97. *Iulii Caesaris Scaligeri Poetices libri septem* (Lyon 1561) 177: "convenientia illa est, quam Graeci συμμετρίαν vocant, Vitruvius commensum."

98. Paul Frankl, *The Gothic: Literary Sources and Interpretations through Eight Centuries* (Princeton: Princeton UP 1960) 88.

99. "Das Baskische und die Sprachwissenschaft," *Hugo Schuchardt-Brevier: Ein Vademecum der allgemeinen Sprachwissenschaft*, ed. Leo Spitzer (Halle: Niemeyer 1928) 228 (first published in *Sitzungsberichte der Akademie der Wissenschaften in Wien* 202/4 [Wien 1925]).

100. For an enlightening discussion of transliteration from the point of view of linguistics, cf. Hans H. Wellisch, *The Conversion of Scripts—Its Nature, History, and Utilization* (New York: Wiley 1978).

101. Paul Valéry, *Tel Quel* (Paris: Gallimard 1943) 2: 50.

102. But also to what they *signify*: that is why a word is always "elegy to what it signifies," as Robert Hass has beautifully written. "Meditation at Lagunitas," *Praise* (New York: Ecco Press 1979) 4.

103. Moritz Haupt's aphorism: "Das Übersetzen ist der Tod des Verständnisses," cit. in Paul Cauer, *Die Kunst des Übersetzens* (Berlin: Weidmann 1914) 4.

104. Benjamin, "Die Aufgabe des Übersetzers," *Gesammelte Schriften* IV·1: 21. Hence the transliteration of the Hebrew text into Greek that made up the second column of Origen's *Hexapla*, his critical edition of the Bible, ought to be regarded as the true "prototype or ideal of all translation."

105. This question is the object of a wonderful story by J. L. Borges, "Averroes's Search," which takes as a pretext the unintelligibility of the word "comedy" in the *Poetics* to somebody "who had never stepped into a theater." Cf. Weinberg, *History of Literary Criticism*, 1: 358–359.

106. Aulus Gellius, *The Attic Nights*, I.xviii; trans. John C. Rolfe, 3 vols. (Cambridge, Mass.: Harvard UP 1946) 1: 87. Cf. Remigio Sabadini, " 'Maccheroni' 'Tradurre' (Per la 'Crusca')," *Rendiconti del R. Istituto Lombardo di Scienze e Lettere*, s. II, 49 (1916) 219–224. Gianfranco Folena does not consider Bruni's *traducere* a misunderstanding, but rather a term of choice (Folena, "'Volgarizzare' e 'tradurre'," 102–103.) Cf. also Lothar Wolf, "Fr. *traduire*, lat. *traducere* und die kulturelle Hegemonie Italiens zur Zeit der Renaissance," *Zeitschrift für romanische Philologie* 87 (1971) 99–105; and Bodo Guthmüller, "Die volgarizzamenti," in *Grundriss der romanischen Literatur des Mittelalters*, vol. X/2, *Die italienische Literatur im Zeitalter Dantes und am Übergang vom Mittelalter zur Renaissance* (Heidelberg: Carl Winter 1989) 201–254.

107. Cf. Panofsky's classic essay, "The History of the Theory of Human Proportions as a Reflection of the History of Styles," *Meaning in the Visual Arts* (Harmondsworth: Penguin 1970), as well as Jürgen Fredel, "Ideale Maße und Proportionen. Der konstruierte Körper," *Die Beredsamkeit des Leibes: Zur Körpersprache in der Kunst*, ed. Ilsebill Barta Fliedl and Christoph Geissmar (Salzburg: Residenz 1992) 11–42.

108. Cf. Panofsky, 276–277.

109. Even if "hoc secundum non satis exprimat Graeci vocabuli significationem." *De verborum Vitruvianorum significatione. Sive perpetuus in M. Vitruvium Pollionem commentarius. Auctore Bernardino Baldo Urbinate.* (Augsburg 1612) 169, *ad vocem* 'symmetria.'

110. Ibidem.

111. Leon Battista Alberti, *L'Architettura [De re aedificatoria]* IX.5, 2 vols. (Milano: Polifilo 1966) 2: 817. He certainly borrowed the term from Cicero's theory of composition. In the *Orator* xxiv.81 Cicero introduced this category to name the quality of an incorrigible composition, so to speak: an arrangement of words that cannot be altered but for the worse. See, for references to other relevant passages in Cicero, Karl Borinski, *Die Antike in Poetik und Kunsttheorie*, vol. 1: *Mittelalter,*

Renaissance, Barock (Leipzig: Dieterich 1914) 149. Valla proposes *concinnitas* as translation of *eurythmia*: "quae concinitas dici potest est venusta species commodusque in compositionibus membrarum aspectus." Spitzer strangely misses the importance of this word for his otherwise exhaustive discussion of *Classical and Christian Ideas of World Harmony* (Baltimore: Johns Hopkins UP 1963; only a footnote on p. 161.)

112. "from the composition of surfaces arises that elegant harmony and grace in bodies, which they call beauty (*ex superficerum compositione illa elegans in corporibus concinnitas et gratia extat, quam pulchritudinem dicunt*) [. . .] the more care and labour they (scil. studious painters) put into studying the proportion of members (*in symmetria membrorum recognoscenda*), the more it helps them to fix in their minds the things they have learned." Alberti, *On Painting*, 73–75.

113. "quidem certa cum ratione concinnitas universarum partium in eo, cuius sint, ita ut addi aut diminui aut immutari possit nihil, quin improbabilius reddatur." Alberti, *Architettura* VI.ii, 2:447. Trans. Leoni, *Ten Books of Architecture*, 113. Cf. also 2: 813: "constat enim corpus omne partibus certis atque suis, ex quibus nimirum si quam ademeris aut maiorem minoremve redegeris aut locis transposueris non decentibus, fiet ut, quod isto in corpore ad formae decentiam congruebat, vitietur." Trans. Leoni, 195: "every Body consists of certain peculiar Parts, of which if you take away any one, or lessen, or enlarge it, or remove it to an improper Place; that which before gave the beauty and Grace to this Body will at once be lamed and spoiled."

114. Alberti, *L'Architettura*, 2: 817.

115. At least in the capacity of a general aesthetic term. See OED *ad vocem*.

116. I.iv. I quote from *Marsilio Ficino's Commentary on Plato's Symposium*, ed. and trans. Sears R. Jayne (Columbia: U of Missouri P 1944) 40; trans. 130. Jayne's translation, whose aim was "litterality," simply gets rid of the problem by erasing all differences in a universal "harmony."

117. The exception being III.iii. (Marsilio Ficino, *El libro dell'amore* [Firenze: Olschki 1987] 52), where "concordia grata" renders the endyadis "rhythmum et concinnitatem" (Jayne 55).

118. Agnolo Firenzuola, *Dialogo delle bellezze delle donne* (1549), *Opere* (Firenze: Sansoni 1971) 538–539. Trans. K.Eisenbichler and J. Murray, *On the Beauty of Women* (Philadelphia: U of Pennsylvania P 1992) 13–14.

119. See *Thesaurus Linguae Latinae, ad vocem*.

120. Cf. Lorenzo de' Medici's *Comento sopra alcuni de' suoi sonetti*: "di più voci concordi resulta un concento, che si chiama 'armonia' " (*Opere*, 2 vols. [Bari: Laterza 1939] 1: 46). See Spitzer, *Classical and Christian Ideas of World Harmony*, chapt. V.

121. *Thesaurus Linguae Latinae, ad vocem.*

122. *Thesaurus Linguae Grecae, ad vocem. Complexio* replaced the transliteration *crasis* in Latin medical literature only in the eleventh century: cf. Danielle Jacquart, "De *crasis* à *complexio*: note sur le vocabulaire du tempèrament en Latin médiéval," *Mémoires V: Textes Médicaux Latins Antiques*, ed. G. Sabbah (Saint-Etienne: Université de Saint-Etienne 1984) 71–76.

123. Cf. Spitzer, *Classical and Christian Ideas of World Harmony*, chapt. IV.

124. Cf. Gravina, writing in the eighteenth century (in Battaglia, *ad vocem*): "*concinnitas*, da *concinendo*, per cagion del suono indi nascente."

125. *Symposium* 187B; trans. Lamb, 5: 127; more than just an *analogy*, or an equality of ratios, from which symmetry is supposed to arise, a *homology*, or an equal ratio. Ludwig Edelstein's article, "The Role of Eryximachus in Plato's *Symposium*," *Transactions of the American Philological Association* 76 (1945) 85–103, does not address this crucial point, which better than any other argument suggests that Eryximachus's speech is not to be taken as a parody of contemporary medical discourse.

126. Giovanni Pico della Mirandola, *Commentary on a Canzone of Benivieni*; trans. Sears Jayne (New York: Lang 1984) 104.

127. Emile Benveniste, *Le vocabulaire des institutions indo-européennes*, 2 vols. (Paris: Minuit 1969) 2: 100–101. See also Thrasybulos Georgiades's *Musik und Rythmus bei den Griechen: Zum Ursprung der abendländischen Musik* (Rowohlt 1958) pp. 91–93, and *Nennen und Erklingen: Die Zeit als Logos* (Göttingen: Vandenhoeck & Ruprecht 1985) 61–62, with relevant quotes.

128. This is implicitly suggested by Lucretius, when he writes that Greek *harmonia* is either the secularization of a divine name, "brought down to musicians from high Helicon," or "perhaps the musicians themselves drew it from some other source and applied it to that which then lacked a name of its own (*proprio quae tum res nomine egebat*)" (*De rerum natura* iii. 132–134; trans. W. H. D. Rouse [Cambridge, Mass.: Harvard UP 1966] 179–181).

129. Cf. Hildebrecht Hommel, *Symmetrie im Spiegel der Antike, Sitzungsberichte der Heidelberger Akademie der Wissenschaften, Philosophisch-historische Klasse* 5 (1986) 22.

130. I allude to Spitzer's above-cited study and to John Hollander's *The Untuning of the Sky: Ideas of Music in English Poetry 1500–1700* (1961; Hamden: Archon 1993).

131. Immanuel Kant, *Kritik der reinen Vernunft*, in *Werke*, 10 vols. (Darmstadt: Wissenschaftliche Buchgesellschaft 1983) 4: 321; trans. Kemp Smith, *Critique of the Pure Reason* (London: Macmillan 1933) 309.

132. Immanuel Kant, "Reflexionen zur Logik," § 3409, *Gesammelte Schriften* (Berlin: Akademie-Verlag 1934) 16: 818.

133. Aristotle, *Problems* XIX.28; trans. W. S. Hett, 2 vols. (Cambridge, Mass.: Harvard UP 1936) 1: 395.

134. He adds Dante's definition of beauty as "harmony," which can also be reduced to Aristotelian beauty, from the *Convivio*; but almost mockingly calls this work "a meager meal" in comparison to Plato's *Symposium*.

135. *Tusc.* IV.v.11: "corporis est quaedam apta figura membrorum cum coloris quadam suavitate eaque dicitur pulchritudo;" trans. J. E. King, *Tusculan Disputations* (Cambridge, Mass.: Harvard UP 1971) 359–360: "in the body a certain symmetrical shape of the limbs combined with a certain charm of colouring is described as beauty." Cicero introduces his definition in the midst of a discussion of the analogy between bodily and soul qualities, which is part of a longer debate on the passions of the soul (cf. also *De Off.* I.91: "pulchritudo corporis apta compositione membrorum movet oculos; delectat hoc ipso, quod inter se omnes partes quodam lepore consentiunt"). Plato had already expressed a similar thought in the *Sophist* 235D (trans. Fowler, 2:333): "I see the likeness-making art as one part of imitation. This is met with, as

a rule, whenever anyone produces the imitation by following the proportions of the original (*kata tas tou paradeigmatos symmetrias*) in length, breadth, and depth, and giving, besides, the appropriate colours to each part." Cf. Creuzer's discussion of the theory Plotinus criticizes, in his edition of *Plotini Liber de Pulcritudine*, ed. Friedrich Creuzer (Heidelberg 1814) 144–154.

136. Benedetto Varchi, *Libro della beltà e grazia* (after 1543): "la bellezza si piglia in due modi, una secondo Aristotele e gli altri che vogliono ch'ella consista nella proporzione de' membri, e questa si chiama et è bellezza corporale [. . .] L'altra bellezza consiste nelle virtù e costumi dell'anima, onde nasce la grazia di che ragioniamo." *Trattati d'arte del Cinquecento*, ed. Paola Barocchi, 3 vols. (Bari: Laterza 1960) 1: 89.

137. "Sunt autem nonnulli qui certam membrorum omnium positionem, sive, ut eorum verbius utamur, commensurationem et proportionem cum quadam colorum suavitate, esse pulchritudinem opinentur." V.iii., p. 67; trans. Jayne, 168.

138. Galen. *de temper.* I 9; trans. J. J. Pollitt, *The Art of Ancient Greece: Sources and Documents* (Cambridge: Cambridge UP 1990) 77. Galen, who is our authority on Polykleitos, concludes that "anyway, it is the opinion of all physicians and philosophers that the beauty of the body resides in the right proportion of the parts." *Stoicorum Veterum Fragmenta* III, ed. von Arnim, fr. 472. This is also the polemical starting point of Plotinus's treatise *On Beauty*: "nearly everyone says that it is good proportion (*symmetria*) of the parts to each other and to the whole, with the addition of good colour, which produces visible beauty, and that with the objects of sight and generally with everything else, being beautiful is being well-proportioned and measured" (*Enneads*; trans. A. H. Armstrong, 6 vols. [Cambridge, Mass.: Harvard UP 1978] 1: 235.)

139. The *homo Vitruvianus* became extremely popular, almost trivial, already by the mid-sixteenth century: cf. Frank Zöllner, *Vitruvs Proportionsfigur: Quellenkritische Studien zur Kunstliteratur im 15. und 16. Jahrhundert* (Worms: Werner 1987).

140. Cf. Plin., *Nat. hist.* XXXIV.55: "fecit et quem canona artifices vocant liniamenta artis ex eo petentes veluti a lege quadam." Trans. H. Rackham, slightly modified, 10 vols. (Cambridge, Mass.: Harvard UP 1952) 9: 169.

141. *Phaedo* 86B: "we believe the soul to be something much like this: our body is as it were tensioned and held together by hot and cold and dry and wet and other things of this sort, and our soul is a blending and *harmonia* of these same things, when they have been finely and proportionately [*metriōs*] blended with one another. So if the soul turns out to be some sort of *harmonia*, it is clear that when our body is excessively [*ametrōs*] slackened or tautened by diseases and other evils, it is inevitable that the soul must perish at once, most divine though it be, just like the other *harmoniai*, those in the notes and in all the things that craftsmen make" (*Greek Musical Writings*, vol. II: *Harmonic and Acoustic Theory*, ed. Andrew Barker [Cambridge: Cambridge UP 1989] 39).

142. "ipsius corporis intentionem quandam, velut in cantu et fidibus quae harmonia dicitur, sic ex corporis totius natura et figura varios motos cieri tamquam in cantu sonos." *Tusc.* I.x.20. Cf. also Macrob. *in somn. Scip.* I.14, 19, and Arist. *de anima* 407b27. According to Walter Burkert, Plato "was the first to point out an

embarassing implication" of the theory, namely, its irreconcilability with the immortality of the soul (*Lore and Science in Ancient Pythagoreanism*, trans. Edwin L. Minar, Jr. [Cambridge, Mass.: Harvard UP 1972] 272).

143. Aristotle rejected such a theory as materialistic in his treatise *On the soul*. Cf. Rohde, *Psyche*, 169, for relevant passages from Macrobius and Claud. Mamert. *de statu animae*, 2.7.

144. Firenzuola, *Bellezze delle donne*, 538–539; trans. Eisenbichler and Murray, 13–14, slightly modified.

145. Etymologies of the name are extremely disparate: from the supposition of an eponym, Physion (Michael Scot, *Liber phisionomie* [Basel ca. 1485] 4ᵛ; cf. also the pseudo-Aristotelian *Secreta secretorum*), to the derivation *physis* and *onoma*, "eo quod ipsam naturam vocat," or *physis* and *nomos*, both attested by Peter of Abano (*Decisiones Physionomiae* [Venetiae 1548] 2v); from the popular etymology <*visum*, which the heading of a XIVth century manuscript in Wolfenbüttel suggests (*De visonomia*, Cod. Guelf. 696 Helmst., 76r), to the seemingly most dependable *physis* and *gnōmē*; although the interpretation (and the spelling, as well: the longer form *physiognōmonikē* being the most ancient one) of the latter half remains controversial: the author of a treatise sometimes attributed to Avicenna interprets *gnomos* as "divinacio" (Roger A. Pack, "Auctoris incerti *De Physiognomonia* libellus," *Archives d'Histoire doctrinale et littéraire du Moyen Age* [1974] 126); Joachim Camerarius interprets *physiognōmonikē* as "notitia et consideratio naturae" (*Commentarius de generibus divinationum, ac Graecis Latinisque earum vocabulis*, Lipsiae 1576] 55); Robert Fludd translates *gnōmē* "signum" (*ΚΑΘΟΛΙΚΟΝ Medicorum ΚΑΤΟΠΤΡΟΝ* [Francofurti 1631] 213), Rudolph Göckel (Goclenius) "regula" (*Physiognomica et Chiromantica specialia* [Francofurti 1625] A [6ʳ]).

146. Firenzuola, *Bellezze delle donne*, 561; trans. Eisenbichler and Murray, 33–34. Jacob Burckhardt rightly singled out this treatise among the wide number of Renaissance writings on beauty and love as particularly interesting from a linguistic point of view.

147. Hobbes, *Works*, 3:14.

148. 41B; trans. Bury, 9: 89.

149. "Longinus," *On the Sublime* xl.1; trans. W. Hamilton Fyfe, revised Donald Russell (Cambridge, Mass.: Harvard UP 1995) 289.

150. Dionysius of Halicarnassus, *On literary composition* 21, *Critical Essays*; trans. Stephen Usher, 2 vols. (Cambridge, Mass.: Harvard UP 1985) 2: 167.

151. F. 3r.

152. German 'harte Fügung,' Norbert von Hellingrath's translation of Dionysius's *austēre harmonia* ("Pindar-Übertragungen von Hölderlin," *Hölderlin-Vermächtnis* [München: Bruckmann 1944] 25).

153. *Ioannis Lygeaei medici de humani corporis harmonia libri IIII* (Lutetiae 1555) f. 4r. Lyege also edited a translation of Hippocrates's aphorisms: *Hippocratis aphorismi: ex Guilielmi Plantii interpretatione, et Ioannis Lygaei annotationibus illustrati: omnia nunc primum ab eodem aucta & emendata* (Genevae 1580).

154. Cf. Gregory Nagy, *The Best of the Achaeans: Concepts of the Hero in Archaic Greek Poetry* (Baltimore: Johns Hopkins UP 1979) 297–300, on the etymology of the name *Homeros* as "he who fits [the song] together."

155. *The Rape of Lucrece*, ll. 1427–1428.

156. Cf. Carlo Valgulio's explanation of the term in the introduction to his translation of Plutarch's *De Musica* (1507): "Melos, in the first meaning, is defined as a member of the body. Although there are many and diverse members in the body of a living being that fulfill various and diverse functions, they all serve the single end of preserving life. Thus, since diverse notes mixed concordantly make a single consensus (*diuersae uoces mixtae concorditer concentum unum conficiunt*), the resultant song the Greeks called melos, and we call it air or melody" (cit. in Palisca, *Humanism in Italian Renaissance Musical Thought*, 93.)

157. Synonymous expressions used in the Homeric poems to refer to the limbs are discussed in H. A. Paraskevaides, *The Use of Synonyms in Homeric Formulaic Diction* (Amsterdam: Hakkert 1984) 66.

158. Diels-Kranz fr. 16. Cf. Lohmann, *Musiké und Logos*, 7.

159. François Lasserre observes apropos of the Hippocratic treatise *De locis in homine* I discussed above, that "l'asyndète [. . .] exprime la volonté de donnner à la description le caractère d'une énumération," "Sociolectes hippocratiques dans le traité *Des Lieux dans l'homme*," *Formes de pensée dans la collection hippocratique; Actes du IV colloque international hippocratique* (Genève: Droz 1983) 167.

160. Camerarius, *Commentarii utriusque Linguae*, c. 40.

161. "Longinus," *On the Sublime* XXI.1; trans. Hamilton Fyfe, 239.

162. "il discorrere è come il correre, e non come il portare, " Galileo Galilei, *Saggiatore*, cit. in Italo Calvino, *Lezioni americane* (Milano: Garzanti 1988) 43. Cf. also Roland Barthes, *Fragments d'un discours amoreux* (Paris: Seuil 1977) 7.

163. The austere style (*austērē harmonia*) wishes "to portray emotion (*pathos*) rather than moral character (*ēthos*)." Dionysius, *On literary composition* 22, 2: 171. See below, chapter 2, note 106.

164. Seneca, *Ad Lucilium Epistulae Morales* XL,1: 271. In Cicero this was still a sign of the superiority of the Greeks: "they thought that in speeches the close of the period ought to come not when we are tired out but where we may take breath, and to be marked not by the punctuation of the copying clerks but by the arrangement of the words and of the thought" (*De oratore*, III.xliv.173, trans. H. Rackham, 3 vols. [Cambridge, Mass.: Harvard UP 1942] 3: 138–139).

165. Seneca, *Epistulae Morales*; trans. Gummere, 1: 271.

Chapter 2

1. Hippocrates, *Prognostic*, I.ii; trans. W. H. S. Jones, 2: 9.

2. "To become another is a kind of death" (*il diventare un altro è una specie di morire*), a character tellingly laments in Giovan Battista Della Porta's comedy *The Astrologer*, voicing a deeply seated resistance to disguise. See *L'Astrologo*, II.ii, in *Le Commedie*, 3 vols. (Bari: Laterza 1911) 2: 325.

3. Cf. Arthur D. Nock, "Cremation and Burial in the Roman Empire," (1932), rpt. in *Essays on Religion and the Ancient World*, 2 vols. (Oxford: Clarendon 1972) 1: 293–294: "mummification guaranteed to the dead man a body which ensured his

continued ability to receive offerings. This was a fundamental aspect of the afterlife, far more intimately related to funerary rites as commonly performed than the theological aspect ever was. The ritual texts professed to help, in fact to enable the dead man, or some part of him, to rise to heaven and sit in the boat of the Sungod; they called him an Osiris, as much as to say that after death he was going through the same triumphant reassertion of self as the traditional dead king [. . .] essentially the rite carried its own guarantee that the individual would keep his full vitality. There was no corporate resurrection to await." In the seventeenth century the Dutch physician Frederik Ruysch enjoyed extraordinary fame because of his newly devised technique of enbalming, by which he was able to prolongate, as Fontenelle puts it, the very lives of his mummies, "au lieu que celles de l'ancienne Égypte ne prolongeaient que la mort" (Fontenelle, "Éloge de Ruysch," *Œuvres complètes* [Genève: Slatkine 1968] 1: 457).

4. A slight variation on Gombrich's apophtegm: "The 'Egyptian' in us can be suppressed, but he can never be quite defeated" (*The Story of Art* [London: Phaidon 1951] 422).

5. Cf. Olivier Reverdin, *La Religion de la Cité Platonicienne* (Paris: Boccard 1945) 107–124; Marcel Piérart, *Platon et la Cité grecque: Théorie et réalité dans la Constitution des "Lois"*, Académie Royale de Belgique: Mémoires de la Classe des Lettres, 2ème série, t. LXII, fasc. 3 (1974) 187–189.

6. "ton tōn sarkōn onkon" (959A–C); trans. R. G. Bury, 10: 533. On the usage of the word *eidōlon* in this context, cf. Jean-Pierre Vernant, "Psuche: Simulacrum of the Body or Image of the Divine?," *Mortals and Immortals: Collected Essays*, ed. Froma I. Zeitlin (Princeton: Princeton UP 1991) 190–191. Bringing to its extreme consequences this line of thought, Philo will justify the unexplained slaying of Er (meaning "leathern" in Hebrew) as God's disposal of the body, "our 'leathern' bulk (*ton dermatikon onkon*)" (Philo, *Legum Allegoria* III.69; trans. G. H. Whitaker, *Allegorical interpretation of Genesis II., III.*, 1: 347).

7. Cf. the discussion of this (*Meteor.* 389b31) and related passages in Heinrich Von Staden, "The Discovery of the Body: Human Dissection and Its Cultural Contexts in Ancient Greece," *Yale Journal of Biology and Medicine* 65 (1992) 232–233.

8. 524B–D; trans. W. R. M. Lamb, 3: 523.

9. Cf. Erich Auerbach's justly famous essay on "Ulysses' Scar" in his *Mimesis. Dargestellte Wirklichkeit in der abendländischen Literatur* (Bern: A. Francke [1946]); trans. Willard Trask, *Mimesis: The Representation of Reality in Western Literature* (Garden City, N.Y.: Doubleday 1957).

10. Samuel Purchas, *Microcosmus or the Historie of Man* (London 1619; rpt. New York: Da Capo 1969) 1.

11. *In Ps.* I.5 (J.-P. Migne, *Patrologia Graeca*, 161 vols. [Paris 1857–1887] 12: 1093), cited by Gisbert Greshake, "Theologiegeschichtliche und systematische Untersuchungen zum Verständnis der Auferstehung," in G. Greshake and Jacob Kremer, *Resurrectio mortuorum: zum theologischen Verständnis der leiblichen Auferstehung* (Darmstadt: Wissenschaftliche Buchgesellschaft 1986) 205.

12. 80C–D. Trans. Fowler, 1: 281.

13. "tam diu durare dicunt quam diu durat et corpus: unde Aegyptii periti sapientia, condita diutius reservant cadavera, scilicet ut anima multo tempore perduret

et corpori sit obnoxia nec cito ad alios transeat;" cited in Nock, "Cremation and Burial," 287. Cf. also Franz Cumont, *Lux Perpetua* (Paris: Geuthner 1949) 16. Saunders argues (*Transition from Ancient Egyptian to Greek Medicine*, 23) that "mummification was regarded as a therapeutic procedure, although transcendentally conceived, which would prevent the physical destruction of the body, in the same way that the physician, by elimination of putrefactive material from the body, could allay disease. Consequently we find the word *sdwh* means not only 'to embalm,' but in medical context 'to treat.' " Majno discards this equivocity as a mere *lapsus calami* (Guido Majno, *The Healing Hand: Man and Wound in the Ancient World* [Cambridge, Mass.: Harvard UP 1975] 127).

14. Cf. also the conclusion of the dialogue (115C–116A), with the friendly rebuttal of Crito's concerns, which almost literally anticipates the passage from the *Laws* I quoted earlier.

15. Cf. the very helpful discussions of the early Christian debate over the resurrection of the flesh by Horacio E. Lona, *Über die Auferstehung des Fleisches: Studien zur frühchristlichen Eschatologie* (Berlin: de Gruyter 1993), vol. 66 of the *Beihefte zur Zeitschrift für die neutestamentliche Wissenschaft und die Kunde der älteren Kirche*; and Greshake, "Untersuchungen," 165–371. See also the first chapter of Carolyne Walker Bynum, *The Resurrection of the Body in Western Christianity, 200–1336* (New York: Columbia UP 1995) 21–58.

16. Origen, *Contra Celsum*, VII.36, cit. in Greshake, "Untersuchungen," 185.

17. Fr. 196. Origen, *Contra Celsum*, V.14; trans. Chadwick, 275.

18. "Resurgere autem non potest nisi quod cecidit," *Adv. Marc.* V.9 (cf. A. H. C. van Eijk, " 'Only that can rise which has previously fallen': The History of a Formula," *Journal of Theological Studies* 22 [1971] 517–529).

19. The same pun is possible in Italian, and Dante has exploited this possibility in one of the most condensed similes of the *Comedy*: "E caddi come corpo morto cade." (*Inf.* V.142).

20. Tertullian, *De resurrectione carnis* 18; trans. Ernest Evans, *Treatise on the Resurrection*, (London: S. P. C. K. 1960) 51. The English "corpse" is the collapse of the Latin body, so to speak.

21. Tertullian, *De resurrectione carnis* 19; trans. Evans, 55. Even when he uses the apparently contradictory simile of the face as "the mirror of all intentions" of the soul (15, trans. 41), he makes sure that it would not be interpreted in spiritual terms: "Never is soul apart from flesh, so long as it is in the flesh: it performs no act without it, for apart from it it does not exist." (ibid.) Tertullian's "energic realism" is vital to Auerbach's recovery of the figural sense in his reading of Dante (Auerbach, "Figura," *Gesammelte Aufsätze zur romanischen Philologie*, 55–92; on Tertullian esp. 65–69). Plotinus quite tellingly concludes from the existence of images to the existence of mirrors: "there would be no image [. . .] if a mirror or something of the sort did not exist. For that whose nature is to come into existence in something else would not come into existence if that something else did not exist, for this is the nature of an image, being in something else" (*Enn.* III.6.14, *On the Impassibility of the Things without Body*; trans. Armstrong, 3: 267).

22. Tertullian, *De resurrectione carnis* 5; trans. Evans, 18–19.

23. Tertullian, *De resurrectione carnis* 34.

24. "adeo vivere totum animae carnis est ut non vivere aliud non sit animae quam a carne divertere," Tertullian, *De resurrectione carnis* 7; trans. Evans, 25.

25. Cf. *De anima* 22, 2: "Definimus animam dei flatu natam, immortalem, corporalem, effigiatam [. . .]" (ed. J. H. Waszink [Amsterdam: Meulenhoff 1947] 31.)

26. Tertullian, *De resurrectione carnis* 17; trans. Evans, 47.

27. Robert Klein, *La forme et l'intelligible* (Paris: Gallimard 1954) 38; trans. Madeline Jay and Leon Wieseltier, *Form and Meaning* (New York: Viking 1979) 67.

28. *Summa contra Gentiles* IV, 90, cited in Greshake, "Untersuchungen," 231.

29. According to the Neoplatonics, too, the soul "declines" toward a body, but is certainly never nostalgic of its former body: cf. e.g., Aristides Quintilianus's account of the descent of the soul in his *De Musica* II.17, in *Greek Musical Writings*, vol. II: *Harmonic and Acoustic Theory*, 488–492.

30. "Just as we are held fast by longings and/by other sentiments, our shades take form" (trans. Allen Mandelbaum, *The Divine Comedy of Dante Alighieri* (Berkeley: U of California P 1982) 2: 235).

31. Cf. Klein's commentary in *La forme et l'intelligible*, 38–40, 107–108.

32. *The Works of Sir Thomas Browne*, 1: 166. The work, probably written in 1656, was first published posthumously in 1690. This passage probably provided the inspiration for Tennyson's lines: "As sometimes in a dead man's face, [. . .] A likeness comes out—to some one of his race" (*In Mem.* lxxiii).

33. *The Works of Sir Thomas Browne*, 1: 170. Vernant uses the image of a "photographic developer" to describe the effects of the "beautiful death" on the bodily appearance of the fallen warrior ("A 'Beautiful Death' and the Disfigured Corpse in Homeric Epic," in Vernant, *Mortals and Immortals*, 50).

34. Hogarth mispells the word as "caracatura" in the inscription to the engraving of *The Bench* (cf. Joseph Burke, "Introduction," in William Hogarth, *The Analysis of Beauty with the rejected passages from the manuscript drafts and autobiographical notes* [Oxford: Oxford UP 1955] lii). Cf. the malicious remark of an early nineteenth-century highbrow critic, Reverend Ferrers, suggesting that "the education of Hogarth was so confined, that it left him deficient in common orthography; in other words, he could not spell." [E. Ferrer], *Clavis Hogarthiana: or, Illustrations of Hogarth: i.e. Hogarth illustrated from passages in authors he never read, and could not understand*, 2nd edition, enlarged and corrected (London: Nichols, 1817) 5.

35. Cf. Werner Hofmann, *Caricature from Leonardo to Picasso*; trans. M. H. L. (New York: Crown 1957) 15. G. A. Mosini is the pseudonym of Mons. Giovanni Massani (cf. Bruno Migliorini, *Parole e storia* [Milano: Rizzoli 1975] 25).

36. As portraiture properly means the action of portraying, so caricature the action of charging, and not its result. Afterward, the word will be metonymically used to refer to the portrait as thus modified.

37. Cited in the *Vocabolario della Crusca* (ed. 1866), *ad vocem* "caricato."

38. Gombrich's felicitous formulation. Cf. E. H. Gombrich, "The Grotesque Heads," *The Heritage of Apelles: Studies in the Art of the Renaissance III* (Ithaca: Cornell UP 1976) 57–75.

39. Thomas Wright, *A History of Caricature and Grotesque in Literature and Art* (1865; New York: Ungar 1968) 2: "art itself, in its earliest form, is caricature; for it is only by that exaggeration of features which belongs to caricature, that unskilful draughtsmen could make themselves understood."

40. Giorgio Agamben, *Stanze: La parola e il fantasma nella cultura occidentale* (Torino: Einaudi 1977; trans. Ronald L. Martinez, *Stanzas: Word and Phantasm in Western Culture* [Minneapolis: U of Minnesota P 1993] 143). Agamben's own proposal is also not wholly satisfactory: he sees the origin of caricature in the prohibition to blazon the entire body, which had been formulated within the emblematic code. But the integrity of the body is never directly threatened by caricature, and it would be hard to point out any influence of caricature on the figurative technique of the emblematists. Deformation and dismemberment are two different categories.

41. E. Kris (with E. H. Gombrich), "The Principles of Caricature," in E. Kris, *Psychoanalytical Explorations in Art* (New York: Schocken 1974) 202. Cf. Gombrich's own reappraisal of his and Kris's thesis in "Magic, Myth and Metaphor: Reflections on Pictorial Satire" (1989), *The Essential Gombrich*, ed. Richard Woodfield (London: Phaidon 1996) 331–353.

42. *The Works of Sir Thomas Browne*, 1: 170.

43. E. H. Gombrich and Ernst Kris, *Caricature* (Harmondsworth: Penguin 1940) 12.

44. On the physiognomical syllogism and Della Porta's physiognomy cf. Patrizia Magli, "The Face and the Soul," *Fragments for a History of the Human Body*, ed. Michael Feher, 3 vols. (New York: Zone 1989) 2: 87–127, esp. 100–105; *Giovan Battista Della Porta nell'Europa del suo tempo*, ed. Maurizio Terzini (Napoli: Guida 1990). Cf. another extremely interesting occurrence of the term in Browne's *Musæum Clausum, or Bibliotheca abscondita*, first published in 1683 in his *Miscellany Tracts*: "Pictures and Draughts in *Caricatura*, of Princes, Cardinals and famous men; wherein, among others, the Painter hath singularly hit the signatures of a Lion and a Fox in the face of Pope Leo the Tenth." *The Works of Sir Thomas Browne*, 5: 138.

45. Giovanni Pietro Bellori, *Le vite de' pittori scultori et architetti moderni* (1672; Genova: Università di Genova s.a.) 102; Carlo Cesare Malvasia, *Felsina Pittrice: Vite de Pittori Bolognesi*, 3 vols. (Bologna 1678) 1: 380.

46. 769b18–22; trans. A. L. Peck, *Generation of animals* (Cambridge, Mass.: Harvard UP 1953) 419.

47. Cf. Urs Dierauer, *Tier und Mensch im Denken der Antike* (Amsterdam: Grüner 1977); Mario Vegetti, "Anatomia e classificazione degli animali nella biologia antica," *Hippocratica: Actes du Colloque hippocratique de Paris (4–9 septembre 1978)* (Paris: CNRS 1980) 469–483.

48. Cf. Lloyd, *Science, Folklore and Ideology*, 22–26.

49. Cf. Isidorus's definition (XX.xvi.7): "Character est ferrum coloratum quo notae pecudibus inuruntur: χαρακτήρ autem Graece, Latine 'forma' dicitur;" from the branding-iron the name passed on to signify the brand. It is, however, from the technical language of coinage that the word was then borrowed to signify any sort of marking: cf. Alfred Körte, "XAPAKTNP," *Hermes* 64 (1929) 69ff., Peter Steinmetz, "Der Zweck der Charaktere Theophrasts," *Annales Universitatis Saraviensis* VIII (1959) 209–246, esp. 224ff. Cf. also J. Hillis Miller. *Ariadne's Thread: Story Lines* (Yale UP 1992) 28–32.

50. Maler Müller's poem "Adams erstes Erwachen und erste selige Nacht," to which Benjamin refers in his essay "Über die Sprache überhaupt und über die Sprache des Menschen," is certainly not unprecedented in suggesting that Adam may be reading, rather than inventing, the names he gives to the animals; this is indeed a topos

in the physiognomical literature (cf. Georg Gustav Fülleborn, *Beiträge zur Geschichte der Philosophie* II. Bd., 8. Stück [Jena 1797] 18) and the main tenet of the theory of the *signatura rerum*. Browne exposes it in the following terms in his *Religio Medici* II.ii: "there are mystically in our faces certaine Characters which carry in them the motto of our Souls, wherein he that cannot read A.B.C. may read our natures. I hold moreover that there is a Phytognomy, or Physiognomy, not only of men, but of Plants and Vegetables; and in every one of them some outward figures which hang as signes or bushes of their inward forms. The finger of God hath left an Inscription upon all his works, not graphicall or composed of Letters, but of their several forms, constitutions, parts and operations, which, aptly joyned together, do make one word that doth express their natures. By these Letters God calls the Stars by their names; and by this Alphabet Adam assigned to every creature a name peculiar to its Nature" (*The Works of Sir Thomas Browne*, 1: 74–75).

51. He points out, in advance of Tertullian, that the name *adam* is first used to refer to the stuff of which man is made in Gen.ii.7, but he draws altogether different conclusions from his allegorical reading: "There are two types of men; the one a heavenly man, the other an earthly. The heavenly man, being made after the image of God, is altogether without part or lot in corruptible and terrestrial substance; but the earthly one was compacted out of the matter scattered here and there, which Moses calls 'clay' " (*Legum allegoria* I. 31; trans. Whitaker, *Allegorical Interpretation of Genesis II., III.*, 1: 167).

52. Philo, *Legum Allegoria* I. 90–92, 1: 207–208.

53. "since images (*eikōn*) do not always correspond to their archetype and pattern (*archetypōi paradeigmati*), but are in many instances unlike it (*anomoioi*), the writer further brought out his meaning by adding 'after the likeness' to the words 'after the image,' thus showing that an accurate cast (*typon*), bearing a clear impression, was intended" (Philo, *De opificio mundi* 71; trans. Whitaker, *On the account of the world's creation given by Moses*, 1: 57).

54. *Purg.* XXIII.31–33; trans. Mandelbaum, 2: 213.

55. Browne, *Hydriotaphia, Urne-Burial or, A Brief Discourse of the Sepulchrall Urnes lately found in Norfolk* (1658) in *The Works of Sir Thomas Browne*, 4: 33.

56. Philo, *Quod deterius potiori insidiari soleat* 152; trans. F. H. Colson, *That the worse is wont to attack the better*, 2: 303.

57. Cf. *The Rape of Lucrece* vv. 807–812: "The light will show charactered in my brow/The story of sweet chastity's decay,/The impious breach of holy wedlock vow./Yea, the illiterate that know not how/To cipher what is writ in learned books/Will quote my loathsome trespass in my looks."

58. *The Works of Sir Thomas Browne*, 1: 170. On the iconographic tradition of Count Ugolino, cf. Frances A. Yates, "Transformations of Dante's Ugolino," *JWCI* 14 (1951) 92–117.

59. 767b6–8; trans. Peck, *Generation of animals*, 401, combined with Peck's own paraphrasis of the passage in his introduction to the Loeb edition of the *Historia Animalium*, 1: xxiv.

60. "monstrosity (*to teras*) is really a sort of deformity (*anaperia*)" (769b30); trans. Peck, 419.

61. Philo, *Quod deterius potiori insidiari soleat* 78; trans. Colson, 2: 255.

62. John Donne, "Preached at a Mariage," May 30, 1621, *The Sermons of John Donne*, ed. George R. Potter and Evelyn M. Simpson, 10 vols. (Berkeley: U of California P 1957) 3: 250.

63. "the man first fashioned was clearly the bloom (*akmē*) of our entire race, and never have his descendants attained the like bloom, forms and faculties ever feebler having been bestowed on each succeeding generation. I have observed the same thing happening in the case of sculpture and painting: the copies (*mimēmata*) are inferior to the originals (*archetypoi*), and what is painted or moulded from the copies still more so, owing to their long distance from the original." *De opificio mundi* 111; trans. Whitaker, 1: 140–141.

64. Augustine, *Confessiones* VII.x.16, VII.ix.13. On the Platonic origins of this expression (*anomoiotētos topos, Pol.* 273D) cf. Pierre Courcelle, *Les Confessions de Saint Augustine dans la tradition littéraire* (Paris: Études augustiniennes 1963) 50–58, 623–640, who also aptly points out Philo's role in mediating Augustine's reading (esp. 52ff.); Robert Javelet, *Image et ressemblance au douzième siècle: De Saint Anselme à Alain de Lille*, 2 vols. (Paris: Letouzey & Ané 1967) 1: 266–285; Charles Dahlberg, *The Literature of Unlikeness* (Hanover: UP of New England 1988), *passim.*

65. Augustine, *Confessiones* VI.iii.4; trans. William Watts (1631), *Confessions*, 2 vols. (Cambridge, Mass.: Harvard UP 1912) 1: 277.

66. Augustine, *Confessiones* VI.iv.6. On Ambrose as a student of Philo, cf. Courcelle, *Les Confessions de Saint Augustine*, 57–58.

67. Based, rather than on a willful misreading of the first word of the Hebrew text, as suggested by Evans in his edition of Tertullian (*Treatise against Praxeas*, ed. and trans. Ernest Evans [London: SPCK 1948] 93; trans. 135, and Evans's note, 209–210), on a textual mend: cf. Pierre Nautin, "Genèse 1, 1–2, de Justin à Origène," in *In Principio: Interprétations des premiers versets de la Genèse* (Études augustiniennes: Paris 1973) 61–94, esp. 83–86. Nautin individuates Tertullian's source in the *Controversy between Jason and Papiscus*, written around 140 by a Jew converted to Christianism. More generally, the text quoted below from the *Evangelium veritatis* suggests a close affinity of the anonymous author to the Valentinian school.

68. Tertullian, *Treatise against Praxeas* 106; trans. 150. Differently from René Braun (*Deus Christianorum: Recherches sur le vocabulaire doctrinal de Tertullien* [2nd ed., Paris: Études augustiniennes 1977] 587–591), I do not consider necessary to interpret the word *persona* in light of Tertullian's trinitary theology to make sense of this difficult passage (*Against Praxeas* XIV.10). I would rather argue that a literal reading is the most consistent with Tertullian's hermeneutics.

69. Cf. on this point Antonio Orbe, *Hacia la primera teologia de la procesion del Verbo: Estudios Valentinianos*, 2 vols. (Romae; apud aedes Universitatis Gregorianae 1958) 1: 155–156.

70. Tertullian, *Treatise against Praxeas* 106; trans. 150.

71. Cf. *The Gospel of Truth*; trans. Harold W. Attridge and George W. MacRae, in *The Nag Hammadi Library in English* (San Francisco: Harper & Row 1988) 49–50: "Now the name of the Father is the Son. It is he who first gave a name to the one who came forth from him, who was himself, and he begot him as a son. He gave him his name which belonged to him; he is the one to whom belongs all that exists around him, the Father. His is the name; his is the Son. It is possible for him to be seen. The

name, however, is invisible because it alone is the mystery of the invisible which comes to ears that are completely filled with it by him. For indeed, the Father's name is not spoken, but it is apparent through a Son. In this way, then, the name is a great thing. Who, therefore, will be able to utter a name for him, the great name, except him alone to whom the name belongs and the sons of the name in whom rested the name of the Father, (who) in turn themselves rested in his name? Since the Father is unengendered, he alone is the one who begot him for him(self) as a name, before he brought forth the aeons, in order that the name of the Father should be over their head as lord, that is the name in truth, which is firm in his command through perfect power. For the name is not from (mere) words, nor does his name consist of appellations, but it is invisible. He gave a name to him alone, since he alone sees him, he alone having the power to give him a name. For he who does not exist has no name. For what name is given to him who does not exist? But the one who exists exists also with his name, and he alone knows it, and alone (knows how) to give him a name. It is the Father. The Son is his name. He did not, therefore, hide it in the thing, but it existed; as for the Son, he alone gave a name. The name, therefore, is that of the Father, as the name of the Father is the Son" (38,7–39,26).

72. Cf. chapt. 3 below, p. 65.

73. See Nicholas of Cusa's definition of *filiatio dei* as *deificatio*, "quae et theosis graece dicitur" in his treatise *De filiatione Dei* (1445) § I. On the post-platonic development of the Platonic thought (*Theaet.* 176B) of the *omoiōsis theōi* cf. Werner Beierwaltes, *Proklos: Grundzüge seiner Metaphysik* (Frankfurt/M.: Klostermann 1979) 294–305 and 385–390.

74. Tertullian, *De Baptismo* V.7, *Homily on Baptism*, ed. and trans. Ernest Evans (London: SPCK 1964) 14. Evans's translation, 15: "the image had its actuality in the <man God> formed, the likeness <becomes actual> in eternity."

75. Soeren Kierkegaard, *Stages on Life's Way* (Princeton: Princeton UP 1940) 192. Leonardo's drawing may be the best illustration of Kierkegaard's analogy. Shakespeare calls the son the father's shadow in Sonnet 37 (cf. Frank Kermode, *Forms of Attention* [Chicago: U of Chicago P 1985] 37), the daughter "the father's image" in *The Rape of Lucrece*, v. 1753. Lucrece's father laments that, after the premature death of his daughter, "I no more can see what once I was" (v. 1764).

76. In spite of D'Arcy Thompson's attempt at translating "the old Hebrew way," namely, teleology, into the vocabulary of "Holism" (cf. D'Arcy Wentworth Thompson, *On Growth and Form*, ed. John Tyler Bonner [1st ed. 1898; Cambridge: Cambridge UP 1961] 3, 265).

77. The final line of the first stanza, ΔΑΙΜΩΝ, in the series of "Urworte. Orphisch," the "primal words" Goethe retrieves out of the vocabulary of ancient Orphism. Goethe, *Selected Poems*; trans. Christopher Middleton (Boston: Suhrkamp 1983) 231. Cf. chapt. 5.

78. "Wie kann das einmal 'Geprägte' sich noch weiter 'entwickeln,' was will 'Prägung' überhaupt besagen, wenn sie nicht irgendeine Zeitlang beharrt, sondern nie stillhaltender Wandel ist?," Georg Simmel, "Über die Karikatur," *Zur Philosophie der Kunst: Philosophische und kunstphilosophische Aufsätze* (Potsdam: Kiepenheuer 1922) 94.

79. Ibid.

80. "Das Zerschellen der Form am Leben" is the title of the essay devoted by Lukács to Kierkegaard's relationship to Regina Olsen in his *Die Seele und die Formen* (Berlin: Fleischel 1911; trans. Anna Bostock, *Soul and Form* [Cambridge, Mass.: MIT 1974]). The essay is meant as an answer to the very Simmelian question: "has the concept of form any meaning seen from the perspective of life?" (*Soul and Form*, 28).

81. The source of this anecdote, usually referred to Zeuxis, is probably Xenophon, *Memorabilia* III.x.2. Cf. Erwin Panofsky, *Idea: Ein Beitrag zur Begriffsgeschichte der älteren Theorie* (2nd ed.; Berlin: Hessling 1960) 7.

82. Malvasia, *Felsina Pittrice*, 1: 380.

83. *Trattati d'arte del Cinquecento*, 1: 209–269.

84. Cf. Panofsky, *Idea*, 59–60.

85. Still commended by Vasari in the "Proemio" to the second part of his *Vite*, as it essentially contributed to the establishment of the third *maniera*. In this, however, *electio* has been transcended by *leggiadria* (see below).

86. In a belated version of the myth of Acteon, Chiron, who initiated Acteon to hunting, has to shape an image of him to calm the howling pack of the dogs, which have just devoured their master "sub imagine cervi" (Ovid, *Metam.*, III, 250). Apollodorus, who records this appendix to the myth, does not specify the nature of this image (*eidōlon*; Apollodorus, *Bibl.*, III, iv, 4). Yet none better than the *centaur* Chiron could be chosen as mythical patron for the amphibious genre of the caricature.

87. Cf. the precise definition given in *The Spectator* no. 537, 4: 417: "the Art consist in preserving, amidst distorted Proportions and aggravated Features, some distinguishing Likeness of the Person."

88. A Platonic association (*Timaeus* 28C); cf. John Onians, "Alberti and ΦΙΛΑΡΕΤΗ: A Study in their Sources," *JWCI* 34 (1971) 108.

89. *Oratio (De hominis dignitate)*, in Giovanni Pico della Mirandola, *De Hominis Dignitate - Heptaplus - De Ente et Uno e scritti vari*, ed. Eugenio Garin (Firenze: Vallecchi 1942) 104–105; trans. Elizabeth Livermore Forbes, *On the Dignity of Man*, in *The Renaissance Philosophy of Man*, ed. Ernst Cassirer, Paul O. Kristeller, John H. Randall, Jr. (Chicago: U of Chicago P 1948) 224–225. On this point cf. Eugenio Garin, *Giovanni Pico della Mirandola: Vita e dottrina* (Firenze: Le Monnier 1937) 200 ff.

90. On God the artist cf. Ernst Kris and Otto Kurz, *Legend, Myth and Magic in the Image of the Artist* (New Haven: Yale UP 1979) 53–60.

91. "If I am not mistaken (*si recte interpretor*), the sculptor's art of achieving likeness is directed to two ends: one is that the image he makes should resemble this particular creature, say a man. They are not concerned to represent the portrait of Socrates or Plato or some known person, believing they have done enough if they have succeeded in making their work like a man, albeit a completely unknown one. The other end is one pursued by those who strive to represent and imitate not simply a man, but the face and entire appearance of the body of one particular man (*vultus totamque corporis faciem imitari exprimereque elaborant*), say Caesar or Cato in this attitude and this dress [. . .]" Alberti, *On Painting and on Sculpture*, 123.

92. Ibid.

93. Cf. Nicholaus of Cusa's characterization of the physiognomist's task in his *De Coniecturis* (1440 ca.): "illi, qui animarum dispositionem per sensibilia inquirunt [. . .] corpus intuentur atque ex eiusdem cum aliis hominibus atque animalibus differentiis et concordantiis spiritus venantur differentiam" (II, x).

94. Cf. Gombrich, "Ideal and Type in Italian Renaissance Painting," *New Light on Old Masters: Studies in the Art of the Renaissance IV* (Chicago: U of Chicago P 1986) 89–124.

95. Martin Kemp has counted at least seven occurrences of such a warning (cf. M. Kemp, " 'Ogni dipintore dipinge se': a Neoplatonic Echo in Leonardo's Art Theory?," *Cultural aspects of the Italian Renaissance: Essays in Honour of Paul Oskar Kristeller*, ed. Cecil H. Clough [Manchester: Manchester UP 1976] 311–323).

96. As Kenneth Clark well put it, Freud's interpretation, in spite of the many factual inaccuracies, remains "beautiful" and "profound" (*Leonardo da Vinci: An Account of his Development as an Artist* [1939; rev. ed., Harmondsworth: Penguin 1967] 137).

97. "It is an extreme defect when painters repeat the same movements, and the same faces (*volti*) and manners of drapery in the same narrative painting and make the greater part of the faces resemble that of their master, which is something at which I have often wondered. For I have known some, who in all their figures seem to have portrayed themselves from life (*ritratto al naturale*), and in these figures are seen the motions and manners of their creator" (Leonardo da Vinci, *Treatise on Painting*, 2 vols.; trans. A. Philip McMahon [Princeton: Princeton 1956] 1: 55).

98. Cf. Carlo Pedretti, *Leonardo da Vinci On Painting: A Lost Book (Libro A) Reassembled from the Codex Vaticanus Urbinas 1270 and from the Codex Leicester* (Berkeley: U of California P 1964) 35: "*Come le figure spesso somigliano alli loro maestri.* Questo accade chè il giudizio nostro è quello che move la mano alle creazioni de' lineamenti d'esse figure per diversi aspetti, in sino a tanto ch'esso si satisfaccia. E perchè esso giudizio è una delle potenze dell'anima nostra, con la quale essa compose la forma del corpo dov'essa abita, secondo il suo volere, onde, avendo co' le mani a rifare un corpo umano, volentieri rifà quel corpo di che essa fu la prima inventrice. E di qui nasce che chi s'innamora volentieri s'innamorano di cose a loro simiglianti;" 53: "*Precetti, che 'l pittore non s'inganni nella elezione della figura in che esso fa l'abito* Debbe il pittore [. . .] riparare con tutto il suo studio di non incorrere nei medesimi mancamenti, nelle figure da lui operate, che nella persona sua si trova. E sappi che questo vizio ti bisogna sommamente pugnare, conciossiachè egli è mancamento ch'è nato insieme col giudizio; perchè lanima, maestra del tuo corpo, è quella che [fe'] il tuo proprio giudizio; e volentieri si diletta nelle opere simili a quella ch'ella operò nel comporre del suo corpo."

99. See above, p. 40.

100. As some interpreters have concluded on the basis of Leonardo's suggestive formulations (see a review of the "psychobiographical approaches" to Leonardo in Bradley Collins, *Leonardo, Psychoanalysis and Art History* [Evanston. Ill.: Northwestern UP 1997]). But see chapt. 5 below, esp. pp. 110–113.

101. David Summers has suggested that Leonardo's judgment is identical with Aristotle's *sensus communis* (cf. his *The Judgement of Sense: Renaissance Naturalism and the Rise of Aesthetics* [Cambridge: Cambridge UP 1987] 71–75).

102. Cf. the introduction. I use the word "persuasion" in Carlo Michelstaedter's sense (cf. his *La persuasione e la rettorica* [1910; Milano: Adelphi 1982]).

103. Shaftesbury, *Second Characters or the Language of Forms*, ed. Benjamin Rand (1914; rpt. New York: Greenwood 1969) 102.

104. Cf. Goethe, *Gedichte*, ed. Erich Trunz, 2 vols. (Frankfurt/M.: Fischer 1964) 2: 404.

105. Heraclitus's saying: *ēthos anthropōi daimōn* is usually mistranslated as "man's character is his fate" (Charles H. Kahn, *The Art and Thought of Heraclitus: An edition of the fragments with translation and commentary* [Cambridge: Cambridge UP 1979] 81). For different approximations to this difficult fragment, see B. Snell, "Die Sprache Heraklits," *Hermes* 61 (1926) 353–381, esp. 363–364; Martin Heidegger, *Über den Humanismus* (Frankfurt: Klostermann 1949) 44–54.

106. As Walter Benjamin has argued in his essay "Schicksal und Charakter" (see below, note 124). The Aristotelian derivation of *ēthos* from *ethos*, habit (*Eth. Nic.*, II.i, *Eth. Eud.*, II.ii., *Magn. Mor.*, I.vi.2), in other words, must be taken literally and not metaphorically, as Schopenhauer would like (*Die Welt als Wille und Vorstellung*, §. 55). As the example of the "free" fall of grave bodies makes clear, no natural tendency can be altered by habit in the Aristotelian cosmos.

107. Shaftesbury, *Second Characters*, 101. Cf. p. 93: "Nothing being more pleasant to human nature from the beginning as this learning, viz. This is this [. . .] Digito monstrari et dicier hic est. Pers. Sat. I, 28." Shaftesbury also refers to the *Poetics*: "so a child delighted (according to Aristotle's Poetics, IV.) Something learnt." When one says a man has a character, Kant writes accordingly in his *Anthropology*, this means either he has one, or none at all.

108. In the "Proemio" to the third part of his *Lives*.

109. In both the transferred sense of "imitations," "copies," and the proper one of artifacts "made against" the rules of Nature or art. In this second sense cf. Filarete, *Treatise on Architecture*; trans. John R. Spencer, 2 vols. (New Haven: Yale UP 1965) 1:11: "there are some men who are beautiful in the face but [who have] more or less deformed and twisted [*contrafatti e storti*] members."

110. The Vitruvian term for the optical refinements that correct the visual impression of the beholder.

111. *Treatise on Painting*, 1: 121.

112. Yet another count of the features of the face. Pliny counted ten or more (*Nat. hist.* VII.i), Johann Valentin Merbitz, in the seventeenth century, conservatively only eight, but he thought that these were more than enough to guarantee the virtually infinite combinatory possibilities with which our daily experience confront us (Johann Valentin Merbitz, *De varietate faciei humanae* [Dresdae 1676] A3r). William Gilpin reduces the number to four, "yet they are capable of receiving so many variations, that no two faces are exactly alike" (*Two Essays: one, on the Author's mode of executing rough sketches; the other, on the Principles on which they are composed* [London 1804] 11).

113. Codice Atlantico 327ᵛ, Leonardo, *Studi di fisiognomica*, ed. Flavio Caroli (Milano: Leonardo 1991) 38 (trans. Walker, slightly modified, in Leonardo, *On Painting*, 120.)

114. Cf. § 85, *Libro A*, p. 71: "That It Is Impossible to Memorize All the Aspects and Changes of the Parts of the Body (*Che gli è impossibile che alcuna memoria riservi tutti gli aspetti e mutazioni delle membra*)."

115. Leonardo consistently uses the expression *mandare a mente*, which most translators render as "to commit to memory." "To commit in one's mind" would be a more faithful translation; the expression, moreover, was already current in the eighteenth century precisely in the same context of artistic training. Francis Grose uses it in recommending an analogous system of mnemonics for the caricaturist who "wishes to delineate any face he may see in a place where it would be improper or impossible to draw it;" to which he adds the remark that this is an expression "school-boys" use, as they "point out the different parts of speech in a Latin sentence." *Rules for Drawing Caricaturas: with an Essay on Comic Painting* (London 1788) 14–15. Already Alberti in his *Treatise of Painting* had introduced a system of mnemonics based upon the model of ancient grammar, as described by Quintilian in his *Institutio oratoria*: "I would have those who begin to learn the art of painting do what I see practised by teachers of writing. They first teach all the signs of the alphabet (*elementorum characteres*) separately, and then how to put syllables together, and then whole words. Our students should follow this method with painting. First they should learn the outlines of surfaces, [which are as it were, the elements of painting (*quasi picturae elementa*)] then the way in which surfaces are joined together, and after that the forms of all the members individually; and they should commit to memory (*memoriae commendent*, Italian version *mandino a mente*) all the differences that can exist in those members, for they are neither few nor insignificant." Alberti, *On Painting*, 97–101; cf. Quintilian, *Institutio oratoria* I.i. 21–31; and a similarly rigorous apprenticeship in the elements of painting will be consistently advocated by most art theoreticians of the following centuries up to Hogarth, whose practice Grose's treatise is meant to illustrate. It is especially important for Quintilian that the children "learn to know the letters from their appearance (*facies*) and not from the order in which they occur. It will be best therefore for children to begin by learning their appearance (*habitus*) and names just as they do with men" (trans. H. E. Butler, 4 vols. [Cambridge, Mass.: Harvard UP 1920] 1: 33).

116. *Treatise on Painting*, 1: 153.

117. See chapt. 3, pp. 64–66. It is important to remark that Leonardo does not provide a rule for the air; he just reminds his pupil that "if you wish to have facility in remembering the air of a face (*in tenerti a mente [la] un aria d'un uolto*), first commit to memory (*inpara prima a mente*) many heads, eyes, noses, mouths, chins, and throats, necks, and shoulders," *Treatise on Painting*, 1: 153. Michael W. Kwakkelstein's conclusion: "according to Leonardo, the typical form of the facial parts defined the *aria* of a face" (*Leonardo da Vinci as a physiognomist: theory and drawing practice* [Leiden: Primavera Pers 1994] 113) is therefore inaccurate. But Leonardo's position is inevitably ambiguous, as he stands at the turning point between the summatory method of ancient portraiture and the new "ideal" approach to the question of portraiture. Kenneth Clark implicitly draws this point when he observes in reference to an early work, the *Annunciazione*, that in this painting "the features are not felt as part of the structure of the face, but are drawn on it." Clark, *Leonardo da Vinci*, 27.

118. *The Literary Works of Leonardo da Vinci*, ed. Jean Paul Richter, 2 vols. (1883; 3rd ed. New York: Phaidon 1970) § 520, 1: 316–317.

119. When we see the face under a favorable light, we would now say—and this is the image Richter's translation suggests: "Of selecting the light which gives

most grace to faces" (*Della eletione dell'aria che dà gratia ai volti*). Yet, as I argue in chapt. 3, the word "aria" must be taken literally, in its meteorological sense, in this sort of theoretical precepts. I believe that this passage may also shed some light on the usage of the term "favour" in Nicholas Hilliard's *Arte of Limning* (ca. 1600), a central category in his doctrine of portraiture (cf. John Pope-Hennessy, "Nicholas Hilliard and Mannerist Art Theory," *JWCI* 6 [1943] 95.) "Favour" means not just "the good proportion" of a face, as Hilliard defines it at first, but rather its pleasant effect upon the viewer, which it might cause even in spite of ill-proportions (cf. Nicholas Hilliard, *A Treatise concerning the Arte of Limning*, ed. R. K. R. Thornton and T. G. S. Cain [Manchester: Carcanet New Press 1981] 75, 81).

120. Vasari (*Vite*, 555) attributes to the "splendor" of Leonardo's own *bellissima aria* the power of cheering all the sad spirits, as if by way of a sympathetic influence (cf. below chapt. 3, pp. 70–71, Ficino's text from the *De vita*).

121. Richter § 488, 1: 304: "Necessaria cosa è al pittore, per essere bon membrificatore nell'attitudine e gesti che far si possono per li nudi, di sapere la notomia de' nerui, ossi, mvscoli e lacerti."

122. The only way to recapture them is by recreating these conditions artificially, in studio, as it were: "When you want to take a portrait do it in dull weather, or as evening falls, making the sitter stand with his back to one of the walls of the court yard."

123. I quote from Roger de Piles's commentary to du Fresnoy's *De Arte Graphica* in Dryden's translation (1750; cited in W. G. Howard, "Ut Pictura Poesis," *PMLA* 24 [1909] 95).

124. Ibid.

125. "der Charakter entfaltet sich sonnenhaft im Glanz seines einzigen Zuges, der keinen andern in seiner Nähe sichtbar bleiben läßt, sondern ihn überblendet." Walter Benjamin, "Fate and Character"; trans. Edmund Jephcott, *Selected Writings*, ed. Marcus Bullock and Michael W. Jennings (Cambridge, Mass.: Harvard UP 1996) 1: 205 (the translation erases the final clause).

126. *Anthropologie in pragmatischer Hinsicht*, in Kant's *Gesammelte Schriften* (Berlin: Königlich Preußische Akademie der Wissenschaften 1907) 7: 293.

127. "Die Unsterblichkeit ist die nothwendige Fortdauer der Persönlichkeit," *Reflexionen zur Metaphysik*, *Gesammelte Schriften* (Berlin: Akademie 1928) 18: 422.

128. *Soliloquy* III,1, *Standard Edition*, ed. Gerd Hemmerich and Wolfram Benda (Stuttgart-Bad Canstatt: Frommann-Holzboog 1981) I,1: 282–283.

129. Cf. Gert Mattenklott's analysis of the fading of character in his *Blindgänger: Physiognomische Essais* (Frankfurt/M.: Suhrkamp 1986) 40: "Vom Individuum wußte man schon zu Zeiten von Clemens Brentano und E. T. A. Hoffmann, daß es auf Sieg nicht spielen kann, aber aussichtsreich auf Gewinn: Zugewinn von Individualität."

130. Benjamin, "Das Kunstwerk im Zeitalter seiner technischen Reproduzierbarkeit" (dritte Fassung), *Gesammelte Schriften* I·2: 492. Benjamin uses the English word "personality" he borrows from the language of the star-system: "Der Film antwortet auf das Einschrumpfen der Aura mit einem künstlerischen Aufbau der 'personality' außerhalb des Ateliers."

131. Lucretius's verse (*De rerum nat.*, III.58) is quoted by Kant in his *Opus posthumum*. Cf. Manfred Sommer, *Identität im Übergang: Kant* (Frankfurt/M.: Suhrkamp 1988) 83.

132. Jerome McGann has elected Hegel's paradox as the motto of his book, *Black Riders: The Visible Language of Modernity* (Princeton: Princeton UP 1993). I submit that we cannot be content to realize that "the Spirit is a Bone" (*Phänomenologie des Geistes.*)

133. The aether, if it is supposed to exist at all, is for Robert Boyle "such a body as will not be made sensibly to move a light feather," hence weaker than the breath (cit. in Steven Shapin and Simon Schaffer, *Leviathan and the Air-Pump: Hobbes, Boyle, and the Experimental Life, including a translation of Thomas Hobbes, Dialogus Physicus de natura aeris* [Princeton: Princeton UP 1985] 184.) Cf. chapt. 4, note 40.

134. *The Tragedy of King Lear*, V.iii.237–238. Pope-Hennessy observes that what commended Bronzino to his Medici patrons was that "he approached the human features as still life. If the ducal physiognomy had to be reproduced in painting and not just in the impassive art of sculpture, this style was the least undignified" (John Pope-Hennessy, *The Portrait in the Renaissance* [Princeton: Princeton UP 1979] 183).

135. Alberti, *On Painting*, 76. Incidentally, this might also help explain why Romance languages speak of "natura morta" and not simply of "natura in riposo," as Mario Praz proposed to translate "still life." In a sense, nature, the "Living Garment" of deity, never rests.

136. Alberti, *On Painting*, 75.

137. On Valverde and Becerra cf. Roberts and Tomlinson, *The Fabric of the Body*, pp. 210–217. I reproduce the illustration from the 1558 Italian version; the first, 1556 edition, in Spanish, was also published in Rome.

138. Countering Donne's evaluation, according to which "Painters have presented to us with some horrour, the *sceleton*, the frame of the bones of a mans body; but the state of a body, in the dissolution of the grave, no pencil can present to us." *Selected Prose* (Harmondsworth: Penguin 1970) 166.

139. Jacques Derrida has pointed out the multiple valences of the Italian term in his *Mémoires d'aveugle: L'autoportrait et autres ruines* (Paris: Réunion des musées nationaux 1990) 10, apropos of the category of *autoritratto*.

140. That is why the skull is no mere prop in Hamlet's meditation upon Yorick's decay, and he needs to carefully scrutinize it, if only to confirm the tragic knowledge that death is, first and foremost, the threat of a loss of personal identity: "Here hung those lips that I have kissed I know not how oft." *Hamlet* V.i. On the other hand, Browne writes that "handsome formed sculls, give some analogy of flesh resemblance," therefore "it is no impossible Physiognomy to conjecture at fleshy appendicies; and after what shape the muscles & carnous parts might hang in their full consistences." *Hydriotaphia, The Works of Sir Thomas Browne*, 4: 33.

141. "to this favor she must come," *Hamlet* V.i. On this term, cf. here above, note 119.

Chapter 3

1. *Plutarchi Chaeronensi [. . .] Opuscula (quae quidem extant) omnia* (Basileae 1530).

2. Plutarque, *Oeuvres morales*, vol. 7:2, ed. R. Klaerr and Y. Vernière (Paris: Les Belles Lettres 1974) 178.

3. "Of unseemly and naughty Bashfulnesse," *The Philosophy, commonly called, The Morals written by the Learned Philosopher Plutarch of Chaeronea. Translated out of Greek into English, and conferred with the Latine Translations and the French, by Philemon Holland, Doctor of Physick* (London 1603) 134. Holland was also the first translator of Pliny's *Natural History* (1601).

4. *Plutarchi Chaeronensis Moralia [. . .] Guilielmo Xylandro Augustano Interprete* (Basileae 1570).

5. *Les Oeuvres Morales et meslees de Plutarque* (Lyon 1579).

6. *Alcuni Opusculetti de le cose morali del divino Plutarco in questa nostra lingua nuovamente tradotti* (Venetia 1549).

7. *Plutarchi [. . .] Vier vnterschiedliche Tractätlein von der zuviel unziemlichen bäwerischen Schamhafftigkeit [. . .] An jetzo allererst in die hochteutsche Sprach gebracht und vbersetzet, durch Danielem Laelium* (Franckfurt am Mayn 1617).

8. *Plutarchs Moralisch-philosophische Werke*, 9 vols.; trans. I. F. S. Kaltwasser (Wien und Prag 1797) 4: 213.

9. *Auserlesene Moralische Schriften von Plutarch*, 4 vols.; trans. Felix Nüscheler (Zürich 1769) 2: 131.

10. Francesco Piccolomini, *Universa philosophia de moribus [. . .] nunc iterum emendatior in lucem edita* (Venetiis 1594; 1st ed. 1583) 177. He enlists them among the "semivirtutes" (124).

11. *Plutarch's Moralia*; trans. Philipp H. De Lacy and Benedict Einarson, 15 vols. (Cambridge, Mass.: Harvard UP 1959) 7: 47.

12. De Lacy and Einarson, "Introduction," 7: 42.

13. Bruno Zucchelli, "Il ΠΕΡΙ ΔΥΣΩΠΙΑΣ di Plutarco," *Maia* XVII (1965) 220–221.

14. Trans. De Lacy and Einarson. Cf. Douglas L. Cairns, *Aidōs: The Psichology and Ethics of Honour and Shame in Ancient Greek Literature* (Oxford: Clarendon Press 1993) 393–398. Phenomena such as blushing or paling were, of course, paramount in establishing the legitimacy of the physiognomical inference from the physical to the mental. On the foundations of Aristotelian physiognomy cf. Lloyd, *Science, Folklore and Ideology*, 22–26.

15. De Lacy and Einarson, "Introduction," 7: 42. Another, more recent English translation recovers the traditional title: "Bashfulness," though the translator glosses the Greek term, transliterated in the text, as "being put out of countenance:" Plutarch, *Selected Essays and Dialogues*; trans. Donald Russell (Oxford: Oxford UP 1993) 226.

16. Cf. the introduction by Jean Sirinelli to the Belles Lettres edition of *Quomodo adulator ab amico internoscatur* (Plutarque, *Oeuvres morales*, vol. 1: 2 [Paris: 1974] 71–73); and a series of articles in the *Cronache Ercolanesi*, assessing the contribution of the Epicurean Philodemus to the description of this character: E. Kondo, "Per l'interpretazione del pensiero filodemeo sulla adulazione nel PHerc. 1457," *CErc* 4 (1974) 43–56; T. Gargiulo, "PHerc. 222: Filodemo sull'adulazione," *CErc* 11 (1981) 103–127; E. Acosta-Méndez, "PHerc. 1089: Filodemo "Sobre la Adulación," *CErc* 13 (1983) 121–138, F. Longo Auricchio, "Sulla concezione filodemea dell'adulazione," *CErc* 16 (1986) 79–91. Already Plato had qualified the rhetoric of the Sophists by the name of "flattery" (*kolakeia*) (cf. *Gorgias* 463A–B; trans. Lamb, 3: 313).

17. Cf. Zucchelli, 223.

18. *The Characters of Theophrastus*; trans. J. M. Edmonds (Cambridge, Mass.: Harvard UP 1953) 47. Jacob Burckhardt, who discusses this passage in his survey of Italian portraiture ("Das Porträt in der italienischen Malerei," *Beiträge zur Kunstgeschichte von Italien* [Basel: Lendorff 1898]), suggests that the same law that, according to Aelian (*Var.hist.* IV.4), applied to portraits exhibited in public spaces, might have extended its validity even to private houses: "daß nämlich laut Staatsbeschluß den Malern, wie den Bildhauern vorgeschrieben gewesen sei, die dargestellten Leute zu veredeln, und eine Geldbuße von tausend Drachmen habe die Verhäßlicher getroffen" (Burckhardt, 145). The most detailed discussion of the emergence of portraiture as a *genre* in the Italian Renaissance, Gottfried Boehm's *Bildnis und Individuum: Über den Ursprung der Porträtmalerei in der italienischen Renaissance* (München: Prestel 1985), summarily downplays the relevance of physiognomy and ancient characterology and theory of temperaments in this development (92–98).

19. 532 B; Holland, 138 (my italics). Holland, with all the other Renaissance translators, reads γραφεί, "painter," instead of γναφεῖ, "fuller," the lection preferred by most later editors.

20. Porphiry, "On the Life of Plotinus and the Order of his Books," in Plotinus, *Enneads*; trans. Armstrong, 1: 3.

21. On this point see especially Eva C. Kuels, *Plato and Greek Painting* (Leiden: Brill 1978); Jean-Pierre Vernant, "The Birth of Images," in *Mortals and Immortals: Collected Essays* ed. Froma I. Zeitlin (Princeton: Princeton UP 1991) 164–192.

22. Cf. e.g. Filippo Baldinucci, *Vocabolario toscano dell' arte del disegno* (1681) in *Opere* (Milano 1809) 3: 98.

23. Porphyry, "Life of Plotinus," 3–4. Scholars have tentatively identified as a portrait of Plotinus a bust excavated in Ostia: see Gisela M.A. Richter, *The Portraits of the Greeks* (London: Phaidon 1965) vol. 3, figs. 2056–2058.

24. Vasari, *Vite*, 118.

25. "cum [. . .] longo post tempore [. . .] Mediolanum venit, ubi tunc eram,[. . .] nihil tamen prius habuit, nihil antiquius, quam ut vultum cerneret, cuius vidisset imaginem [. . .] quod neque pictor primus votum eius implesset, et mutata annis esset effigies mea, alterum adhibuit, unum quidem ex paucissimis nostri aevi pictoribus adhibiturus Zeuxim aut Prothogenem, aut Parrhasium, aut Apellem si nostro saeculo dati essent, sed omnis aetas contenta suis ingeniis sit oportet, misit ergo quem potuit, magnum prorsus artificem, ut res sunt, qui cum ad me venisset, dissimulato proposito, meque lectioni intento, ille suo iure assidens, erat enim mihi familiarissimus, nescio quid furtim stylo ageret, intellexi fraudem amicissimam, passusque sum nolens, ut ex professo me pingeret, quod nec tamen omni artis ope quivit efficere. sic mihi, sic aliis visum erat, cur si quaeris nescio, nisi quod saepe vehementius tentata succedunt segnius, et nimia voluntas effectum necat" (*Senilia* I. 6; trans. Aldo S. Bernardo, Saul Levin, and Reta A. Bernardo, *Letters of Old Age*, 2 vols. [Baltimore: Johns Hopkins UP 1992] 1: 28–29.) Cf. on this passage E. H. Gombrich in his "Giotto's Portrait of Dante?," *New Light on Old Masters: Studies in the Art of the Renaissance IV* (Chicago: U of Chicago P 1986) 29–30.

26. Cited in Edward J. Martin, *A History of the Iconoclastic Controversy* (1930; New York: AMS 1978) 137. It is not one of the minor ironies of the pluricentennial

controversy over the icons that this passage could become "the *locus classicus* of the orthodox side" (ibid.) Cf. also Michael Psellus's *Commentary on the Chaldaean Oracles*: "Among the philosophers, one calls image what is connatural to superior beings, although it happens to be inferior to them; thus the intellect (*nous*) being connatural to God [. . .] is also its image (*eidōlon tou theou*)" (*Oracles chaldaïques avec un choix de commentaires anciens*, ed. Edouard des Places [Paris: Belles Lettres 1971] 162).

27. First attested in Cennini, and still used in Dolce's time; cf. his "Dialogo della Pittura di M. Lodovico Dolce," (1557) ed. and trans. Mark W. Roskill, in his *Dolce's "Aretino" and Venetian Art Theory of the Cinquecento* (New York: New York UP 1968) 144.

28. The word "inprentare," which most of the early Italian editors resolved in "improntare," before Thompson's diplomatic transcription, suggests more than just "casting:" for it unites in itself the meanings of both French "empreinte" and "emprunt," English "print" and "loan." In a sense, a portrait-painter is also taking over his sitter's identity.

29. Cf. Cennino Cennini, *Il libro dell'arte*, 2 vols., ed. Daniel V. Thompson (New Haven, Yale UP 1932) 1: 117: "Ti voglio tochare d'un'altra, la quale e molto utile, e al disengnio fatti grande honore, in ritrarre e simigliare chose di natural; la quale si chiama inprentare."

30. G. Leopardi, *Zibaldone di pensieri* [170], in *Tutte le opere*, 2 vols. (Firenze: Sansoni 1969) 2: 81.

31. See on this especially E. H. Gombrich, "Leonardo on the Science of Painting: Towards a Commentary on the 'Trattato della Pittura'," *New Light on Old Masters*, 32–60.

32. The analogy, which is already in Seneca (*Epist.* LXXV: "similem esse te volo, quomodo filium; non quomodo imaginem"), is taken up by Paolo Cortese in his letter to Poliziano and criticised by Giovan Francesco Pico in his epistolary with Bembo.

33. "curandum imitatori ut quod scribit simile non idem sit, eamque similitudinem talem esse oportere, non qualis est imaginis ad eum cuius imago est, quae quo similior eo maior laus artificis, sed qualis filii ad patrem. In quibus cum magna saepe diversitas sit membrorum, umbra quaedam et quem pictores nostri aerem vocant, qui in vultu inque oculis maxime cernitur, similitudinem illam facit, quae statim viso filio, patris in memoriam nos reducat, cum tamen si res ad mensuram redeat, omnia sint diversa; sed est ibi nescio quid occultum quod hanc habeat vim" (trans. E. H. Gombrich, in his "The Style *all'antica* : Imitation and Assimilation," *Norm and Form: Studies in the Art of the Renaissance I* [London: Phaidon 1971] 122).

34. The Milanese physician Ludovico Settala spells in most clear terms the distinction between "proportio" and "similitudo," or quantitative and qualitative analogy, in his treatise *De naevis* (Mediolani 1606) 43: "ἀναλογίαν, sive conspirationem, ut ita dicam, partium, bifariam solemus attendere; aut quantitatem, sive mensuram, aut substantiae qualitatem, seu conditionem [. . .] alteramque usurpato Proportionis nomine, alteram similitudinis appellatione ob maiorem claritatem affecturus [. . .]." Cf. chapt. 1.

35. Literally, "the body and the face": *to sōma kai to prosōpon* (Plutarch, *Cimon* II. 3 in Plutarch's *Lives*, 11 vols.; trans. Bernadotte Perrin [Cambridge, Mass.: Harvard UP 1928] 2: 409).

36. Trans. J. Pollitt, cited in D. Summers, *Michelangelo and the Language of Art* (Princeton: Princeton UP 1981) 474. Summers suggests that the correspondence between Petrarch's definition of *aria* and Plutarch's definition of his own historiographical method "may reflect the beginning of the revival of interest in Plutarch at Avignon in the latter half of the fourteenth century," (ibid.) which will culminate in Amyot's classic translation. Cf. also his "Aria II: The Union of Image and Artist as an Aesthetic Ideal in Renaissance Art," *Artibus et Historiae* 20 (1989) 26. The importance of representing the eyes as the proper way to express the affections of the mind is first stressed by Socrates in his conversation with Parrhasius, as reported by Xenophon (*Memorabilia*, X.4). Cf. also Plotinus, *Enn.* II.3.7; trans. Armstrong, *On whether the stars are causes*, 2: 69: "because of the one principle in a single living being, by studying one member we can learn something else about a different one. For instance, we can come to conclusions about someone's character, and also about the dangers that beset him and the precaution to be taken, by looking at his eyes (καὶ ἦθος ἄν τις γνοίη εἰς ὀφθαλμούς τινος ἰδών) or some other parts of the body."

37. "Air," according to the definition proposed by Mr. A. Boyer in his *Dictionnaire Royal, François-Anglais et Anglois-François* (1699; London 1756) is "ce qui résulte de tous les traits du visage, harmonie des parties du visage, *principalement dans un tableau*" (my italics), and he gives the example: "Je vois tous les traits, etc. de votre visage dans ce portrait; mais l'air n'y est pas."

38. *Plutarch's Morals. Translated from the Greek by Several Hands* (1684; 5th ed., London 1718) 2: 108. Cf. the translation by Frank Cole Babbitt for the Loeb Classical Library: "bad painters, who by reason of incompetence are unable to attain to the beautiful, depend upon wrinkles, moles, and scars to bring out their resemblances" (53 D). "Features" and "Air" are obviously synonimous, for Tullie (but see below Addison's remark, p. 70). Both Giovan Francesco Pico della Mirandola, in his correspondence on imitation with Pietro Bembo, and Erasmus, in his *Ciceronianus*, resort to this argument in their polemic against Ciceronianism: "Sunt enim qui nevos, qui cicatrices, qui maciem, qui excrementum etiam effingere velint, vel nulla vel minima ratione habita et lacertorum, et vividi roboris, et gratiae" (*Le Epistole "De Imitatione" di Giovan Francesco Pico della Mirandola e di Pietro Bembo*, ed. G. Santangelo [Firenze: Olschki 1954] 29).

39. "The use of the word *air* in the meaning 'manner, appearance' has been explained by all etymologists from Diez to Dauzat [. . .] by reference to Old Fr. *aire* (="aerie"), which is supposed to have telescoped with *air* (Littré, von Wartburg)" (Spitzer, "Milieu and Ambiance," *Essays in Historical Semantics*, 258–259). The same derivation has been most recently, although cautiously, endorsed by Max Pfister in his *Lessico Etimologico Italiano*, fasc. 6, vol. I (1982), c. 1087.

40. Brought forward by etymologists who are "imbued," in his words, "with the spirit of *Wörter und Sachen*, so destructive of the things of the spirit" (Spitzer, 258). Another example of this still pervasive spirit is Jean Renson's study *Les dénominations du visage en français et dans les autres langues romanes*, 2 vols. (Paris: Les Belles Lettres 1962).

41. "*air* ='manière' is none other than *air* = 'atmosphere' " (Spitzer, 259). Similar conclusions had reached Gerhard Rohlfs already in his dissertation *Ager, Area, Atrium: Eine Studie zur romanischen Wortgeschichte* (Berlin 1920) 52–53; cf. also his

reviews of von Wartburg's *Französisches etymologisches Wörterbuch* in the *Literaturblatt für germanische und romanische Philologie* 43 (1922) c. 244, and 45 (1924) c. 227; and Américo Castro, "Adiciones hispánicas al Diccionario de Meyer-Lübke," *Revista de filología española* 5 (1918) 26–28..

42. Spitzer's term for the condensation of two divergent etymological lines.

43. "Tales splendores Graeci areas vocavere quia fere terendis frugibus destinata loca rotunda sunt" (*Naturales Quaestiones* I, 2.3; trans. Thomas H. Corcoran [Cambridge, Mass.: Harvard UP 1971] 25).

44. On the history of the word cf. Gianfranco Folena, "*Chiaroscuro* leonardesco," in *Il linguaggio del caos: studi sul plurilinguismo rinascimentale* (Torino: Bollati Boringhieri 1991) 242–254; and René Verbraeken, *Clair-Obscur,—histoire d'un mot* (Nogent-le-Roi: L.A.M.E. 1979), esp. pp. 83–100, "Les origines italiennes du terme de clair-obscur."

45. See E. H. Gombrich, "The Heritage of Apelles," in *The Heritage of Apelles: Studies in the Art of the Renaissance III* (Ithaca: Cornell UP 1976) 3–18. In explaining such phenomena as the halo, the rainbow, the chasm, Aristotle deployed an array of optical arguments that could not fail to attract the attention of ancient painters. Such are, for instance, the principles he lays down introducing his discussion of the rainbow, the third of which had most important consequences for the practice of art in the ancient world and throughout the Middle Ages: "dark colour is a kind of negation of vision, the appearance of darkness being due to the failure of our sight; hence objects seen at a distance appear darker because our sight fails to reach them." *Meteorologica* 374b12–15 (trans. H. D. P. Lee [Cambridge, Mass.: Harvard UP 1952] 259). It was in commenting upon passages such as this that Philoponos could then compare the illusionistic effects aimed at by painting with the analogous working of light and shadow in nature: "If you put white and black upon the same surface and then look at it from a distance, the white will always seem much nearer and the black further off. Hence when painters want something to look hollow, such as a well, a cistern, a ditch, or a cave, they colour it black or brown. But when they want something to look prominent, such as the breasts of a girl, an outstretched hand, or the legs of a horse, they lay black on the adjoining areas in order that these will seem to recede and the parts between them will seem to come forward." Trans. Gombrich, "The Heritage of Apelles," 5. The passage refers to 342b14, where Aristotle discusses the apparent depth of chasms, due to the breaking out of light from a dark background (*Meteorologica*, 37). Strangely enough, Gombrich limits himself to refer to Philoponos and Longinus (XVII.2) as the sources of this counterintuitive theory.

46. "quel che si dice in una donna 'ella ha aria' non è altro che lo avere un certo buon segno manifestante la sanità dell'animo e la chiarezza della lor conscienzia: con ciò sia che dicendo aria semplicemente, per figura di antonomasia, che noi per eccellenza forse propriamente diremo, e' s'intende della buona. E la mal'aria, e non avere aria, importa un segno, un piglio dimostrante la malattia del cuore, e le macerie della contaminata conscienzia." Firenzuola, *Delle bellezze delle donne*; trans. Eisenbichler and Murray, 40.

47. *Coelestis Physiognomoniae Libri sex. Ioan. Baptistae Portae Neapolitani* (Napoli 1603) 1–2. Differently from Petrarch, Della Porta transliterates the Italian form into Latin instead of translating it into "aerem." Agostino Mascardi, who resorts

to the category of "aria" in his treatise *Dell'arte historica*, in order to define the even more elusive category of "style," stresses the normality of its usage in ordinary language: "quella cosa, che vulgarmente nomiamo aria del volto, è una qualità propria, & individual di ciascuno, nascente dalla particolar complessione, per cui si rende differente dagli altri, co' quali ha le parti con le misure, e con l'ordine, i colori con la loro temperatura communi: e questa da noi per avventura, anzi dal vulgo intesa con l'intelletto, non sappiamo con tutto ciò diffinirla, & esprimerla" (Agostino Mascardi, *Dell'arte historica* [Venetia 1655] 445).

48. "è da sapere che in qualunque parte l'anima più adopera del suo officio, che a quella più fissamente intende ad adornare, e più sottilmente quivi adopera. Onde vedemo che ne la faccia de l'uomo, là dove fa più del suo officio, che in alcuna parte di fuori, tanto sottilmente intende, che, per sottigliarsi quivi tanto quanto ne la sua materia puote, nullo viso ad altro viso è simile; perché l'ultima potenza de la materia, la quale è in tutti quasi dissimile, quivi si riduce in atto. E però che ne la faccia massimamente in due luoghi opera l'anima—però che in quelli due luoghi quasi tutte e tre le nature de l'anima hanno giurisdizione—cioè ne li occhi e ne la bocca, quelli massimamente adorna e quivi pone lo 'ntento tutto a fare bello, se puote [. . .] Li quali due luoghi, per bella similitudine, si possono appellare balconi de la donna che nel dificio del corpo abita, cioè l'anima; però che quivi, avvegna che quasi velata, spesse volte si dimostra. Dimostrasi ne li occhi tanto manifesta, che conoscer si può la sua presente passione, chi bene là mira [. . .] Dimostrasi ne la bocca, quasi come colore dopo vetro" (Dante Alighieri, *Convivio* III. 8, *Opere minori*, eds. Cesare Vasoli and Domenico de Robertis [Milano: Ricciardi 1979] I.2: 389–392; trans. W. W. Jackson, *Dante's Convivio*, slightly modified [Oxford: Oxford UP 1909] 153–154).

49. "Sopra di ogni altra cosa, state di continuo fermi e intenti riguardando ne gli occhi: percioche quivi siedono, quasi tutte le significazioni de le nostre voglie: Et per loro (come per finestre aperte) traspare e traluce l'anima nostra" (Antonio Pellegrini, *Della Fisonomia Naturale* [Milano 1621] 304). The first edition was published in Venezia in 1545 under the title *I segni de la natura ne l'Huomo*. For an earlier example cf. the passage from Galeottus Martius's *De homine* (1490), quoted by Michael Baxandall in his *Painting and Experience in Fifteenth Century Italy* (Oxford: Oxford UP 1972) 58. The sources of the simile in ancient literature are discussed exhaustively in Lactantius, *L'ouvrage du Dieu créateur* (*De opificio dei*), ed. Michel Perrin, 2 vols. (Paris: Éditions du Cerf 1974) 2: 313–315.

50. *De Coloribvs Ocvlorvm* (Florentiae 1550).

51. "cum nam cordis motus aliae omnes partes aliquatenus repraesentent, multo tamen magis id faciunt oculi, & vultus. Maioribus nam venis & arterijs vultus cutem texit, quam reliqui corporis." Io. Baptista Persona, *Noctes Solitariae sive de iis quae scientifice scripta sunt ab Homero in Odyssea* (Venetiis 1613) 42.

52. John Bulwer, *Pathomyotomia or a Dissection of the significative Muscles of the Affections of the Minde. Being an Essay to a new Method of observing the most Important movings of the Muscles of the Head, as they are the neerest and Immediate Organs of the Voluntarie or Impetuous motions of the Mind* (London 1649).

53. James Parsons, *Human Physiognomy Explaind: in the Crounian Lectures on Muscular Motion* (London 1747).

54. "La peau du visage est d'une constitution particulière, qui ne se trouve point ailleurs. Partout, la peau est séparée de la chair: sur le Visage, l'une et l'autre

sont tellement unies, qu'on ne peut les séparer sans les déchirer; ce qui rend la peau du Visage en quelque façon transparente, et plus propre à recevoir les diverses couleurs qui sont excitées par les différents mouvements qui arrivent, et à nous les peindre au dehors." [Jacques Pernetti], *Lettres philosophiques sur les Physionomies* (La Haie 1746) 184–185.

55. Della Porta, *Coelestis Physiognomoniae*, 2.

56. On this issue cf. André Grabar, *L'Iconoclasme byzantin* (Paris: Flammarion 1984); David Freedberg, *The Power of Images: Studies in the History and Theory of Response* (Chicago: U of Chicago P 1989) 205–212; Georges Didi-Huberman, *Devant l'image* (Paris: Editions de Minuit 1990) 221–247; Moshe Barash, *Icon: Studies in the History of an Idea* (New York: New York UP 1992), Christoph Geissmar, "Das wahre Bild. Modelle zur Simulation Christi," *Die Beredsamkeit des Leibes: Zur Körpersprache in der Kunst*, ed. Ilsebill Barta Fiedl and Christoph Geissmar (Salzburg: Residenz Verlag 1992) 43–54.

57. "pictorem eximium ad eum misit, eique ut diligenter & accurate faciem eius in pictura effingeret, imaginemque pro desiderato illo sibi afferret, imperavit. Et ille quidem venit, atque in loco sublimiore stans, pingere Christi vultum, ut decebat, conabatur. Quum vero coeptum opus non succederet (diuinus quippe splendor & gratia in vultu eius coruscans impedimento erat) re ea cognita SALVATOR textum lineum petit, in eoque prius loto faciem suam exprimit, & ad Abgarum mittit" (ΝΙΚΗΦΟΡΟΥ ΚΑΛΛΙΣΤΟΥ ΕΚΚΛΗΣΙΑΣΤΙΚΗΣ ΙΣΤΟΡΙΑΣ [. . .] Latina interpretatio Ioannis Langi [Lutetiae Parisiorum 1630] 1: 145). I quote William Caxton's version of this passage from his translation of the *Legenda Aurea* (London 1483) 339, c. 2. Della Porta limits himself to relate the failure of the painter.

58. *Lives*, 9: 7.

59. Erasmus, *Ciceronianus*, in *Ausgewählte Schriften*, 8 vols. (Darmstadt: Wissenschaftliche Buchgesellschaft) 7: 108. On Erasmus's attitude toward images and the visual arts, see Erwin Panofsky, "Erasmus and the Visual Arts", *Journal of the Courtauld and Warburg Institute* 32 (1969), 200–227; Helmut Feld, *Der Ikonoklasmus des Westens* (Leiden: Brill 1990) 110–115.

60. Cf. Sir Joshua Reynolds, *Portraits*, ed. Frederick W. Hilles (New York: McGraw-Hill 1952) 74: "The habits of my profession unluckily extend to the consideration of so much only of character as lies on the surface, as is expressed in the lineaments of the countenance."

61. *The Spectator*, 1: 366, 368 (June 8, 1711, No. 86).

62. Baxandall, *Giotto and the Orators*, 26.

63. Such as Girolamo Manfredi's *Opera nova intitulata Il Perche utilissima ad intendere la cagione de molte cose: & maximamente alla conservatione della sanita: Et phisionomia. Et virtu delle herbe* (Ancona 1512; first Latin edition, *Liber de homine*, Bologna 1474): "Come dice Ptholomeo che le face e figure de questo mondo sono subiecte ale face e figure del cielo" (f. 75r). On Manfredi as popularizer cf. Charles Singer, "A Study in Early Renaissance Anatomy, with a new text: The *Anothomia* of Hieronymo Manfredi (1490)," *Studies in the History and Method of Science*, ed. C. Singer (Oxford: Oxford UP 1917) 79–164.

64. Such as Francesco Giorgio Veneto: "si vera est Ptolomaei doctrina, quod aspectus huius terrestris mundi coelestibus vultibus subiiciantur, ut sit Scorpio terrestris sub scorpione coelesti: et taurus, aut aries hic degens sub coelestibus imaginibus: multo magis homo, qui est totus mundus, convenit cum omnibus coelestibus aspectibus"

(*De harmonia mundi totius Cantica tria* [Paris 1544] 111r). Cf. Ps.-Ptolemy, *Centiloquium*, aph. 9 (Venice 1484); Ficino, *Three Books on Life*; trans. Kaske and Clark, 305: "Ptolemy says in the *Centiloquium* that images of things here below are subject to the celestial images; and that the ancient wise men used to manufacture certain images when the planets were entering similar faces of the heavens, the faces being as it were exemplars of things below (*Ptolomaeus ait in* Centiloquio *rerum inferiorum effigies vultibus coelestibus esse subiectas, antiquosque sapientes solitos certas tunc imagines fabricare, quando planetae similes in coelo facies quasi exemplaria inferiorum ingrediebantur*)."

65. "Nonne principis in urbe vultus quidem clemens et hilaris exhilarat omnes? Ferox vero vel tristis repente perterret? Quid ergo coelestium vultus, dominus omnium terrenorum, adversus haec efficere posse putas? Quippe cum etiam coeuntes ad prolem plerunque vultus, non solum quales ipsi tunc agunt, sed etiam quales imaginantur, soleant filiis diu postea nascituris imprimere, vultus eadem ratione coelestes materias confestim suis notis inficiunt, in quibus si quando diu latitare videntur, temporibus deinde suis emergunt" *Three Books on Life*, l. III c. XVII: "Quam vim habeant figurae in coelo atque sub coelo"; trans. Kaske and Clark, 331.

66. Cf. Gundel, *Dekane und Dekansternbilder*, 34–36.

67. *Three Books on Life*, 331.

68. "Facies magis ad corpus: Vultus magis ad animum refertur, atque voluntatem, unde descendit. Nam volo supinum habebat vultum: inde dicimus irato et moesto vultu potius quam facie: et contra lata aut longa facie, non vultu" (Lorenzo Valla, *Elegantiarum Libri VI*, in *Opera* [Basileae 1540] 125, quoted by Baxandall, *Giotto and the Orators*, 172).

69. Cicero, *De legibus* I.9; trans. Clinton Walker Keyes (Cambridge, Mass.: Harvard UP 1928) 327: "the countenance, as we Romans call it, which can be found in no living thing save man, reveals the character. (The Greeks are familiar with the meaning which this word 'countenance' conveys, though they have no name for it.)" (*is qui appellatur vultus, qui nullo in animantem esse praeter hominem potest, indicat mores: cuius vim Graeci norunt, nomen omnino non habent*). Cf. also *Oratio pro Cluentio* 72: "recordamini faciem atque illos eius fictos simulatosque vultus."

70. Gerardi Joannis Vossii *Etymologicon Linguae Latinae* (Amstelodami 1662), *ad vocem*.

71. This view is not shared by more recent etymologists, who interpret it as "was (dem andern) gegen die Augen gerichtet ist" (Eduard Schwyzer, *Griechische Grammatik* [München: Beck 1950] 2: 517), therefore more in the vein of *vultus a volvendo* (see below.) Cf. Françoise Frontisi-Ducroux, *Du masque au visage: Aspects de l'identité en Grèce ancienne* (Paris: Flammarion 1995) 19–38.

72. On the whole issue, see L. Malten, "Die Sprache des menschlichen Antlitzes in der Antike," *Forschungen und Fortschritte* 27 (1953) 24–28.

73. Cf. a standard lexicon of the Baroque era, M. Martinius's *Lexicon Philologicum, Praecipue Etymologicum* (Bremae 1623) *ad vocem* : "*Vultus*, habitus faciei, qui indicat, quid volimus: a volo. Donat. in Andr. I.i. ad illud *Forma et vultu*, annot. *Forma immobilis est et naturalis: vultus movetur et fingitur*. Ergo et a volvendo dici potest." Tertullian suggests another possible derivation in the same spirit in *De resurrectione* 15, from the inchoative *volutare*, "to turn around over and over," hence

"to brood over": "volutet aliquid anima, vultus operatur indicium" (trans. Evans: "Let the soul consider a matter: the countenance tells the tale," 41).

74. "quo vero ad verba attinet, ille qui phisiognomiçat, graece dicitur physiognomon, placet ergo mihi novo nomine appellare uultispectorem et ipsum physiognomiçare, uultispicere. nam uultus superficies est, quae cernitur, a volvendo dictus. Non enim pro facie tantum, sed pro omni superficie qui apparet, accipitur" (Agostino Nifo, *Parva Naturalia [. . .], videlicet Physiognomicorum libri tres [. . .] Omnia post primas editiones nunc primum emendatiora in lucem prodeunt* [Venetiis 1550] f. 3r.) *Vultus a volvendo* is probably the most likely derivation, if one considers the most credited etymology of *prosōpon* (see note 71, above): that which is turned toward somebody else. *Vultus a volendo* is probably a later derivation introduced to distinguish the face in itself from the face as related to an observer, cf. German *an sich* and *für sich*.

75. "Aliud est facies, aliud vultus: illa naturae est ac immutabilis; hic voluntatis, eoque mutabilis" (Vossius).

76. Cf. Roberto Weiss, "Per la storia degli studi greci del Petrarca: il *Triglossos*," *Annali della SNS di Pisa: Lettere, Storia e Filosofia*, serie II, vol. XXI (1952), 254; and Mariarosa Cortesi, "Petrarca, il Triglossos e il Penthaglossos," *Studi petrarcheschi* n.s. 6 (1989), 207, who transcribe differently the very corrupt manuscript (Weiss: "...effigiem faciem dat"; Cortesi: "...effigies faciem dat"); I believe that my interpretation re-establishes the correct reading of the line.

77. Cf. Isidorus's etymology, which simply reverses the relationship between the two terms: "Facies dicta ab effigie" (*Etymologiae* XI, i, 33).

78. Cennini, *Il libro dell'arte*, 1: 120 (trans. Thompson, 2: 127).

79. Max J. Friedländer, *Landscape, Portrait, Still-Life: Their Origin and Development* (New York: Schocken 1963) 236. Cf. Leonardo, *Libro A*, ed. Pedretti, 42–43: "Negli atti affezionati dimostrativi di cose propinque per tempo o per sito, s'hanno a dimostrare con la mano non troppo remota da essi dimostratori; e se le predette cose saranno remote, remota debbe essere ancora la mano del dimostratore e *la faccia del viso volta a chi si dimostra*." (my italics)

80. Along with Portuguese: cf. the particularly interesting instance from J. de Deus's "Campo de flores," quoted in Renson, 2: 641: "Vulto aereo". But the usage of the term even in literary Portuguese is apparently far less frequent than in Italian (ibid., 640).

81. Spanish "bulto" <*vultus* has come to mean the "bulk" of a person or an object, "todo aquello che hace cuerpo," according to the precious definition of the *Diccionario de Autoridades, ad vocem*.

82. "Who is she who comes, that everyone looks at her,/Who makes the air tremble with clarity" (trans. Lowry Nelson, Jr., *The Poetry of Guido Cavalcanti* [New York: Garland 1986] 7). Domenico De Robertis, in his edition of Cavalcanti's *Rime*, aptly points out the connection between these two lines and the optical theory of "scintillazione" (Guido Cavalcanti, *Rime con le rime di Iacopo Cavalcanti*, ed. D. De Robertis [Torino: Einaudi 1986] 17).

83. *Opera de l'antica, et Honorata scientia de Nomandia, specchio d'infiniti beni, et mali, che sotto il cerchio della Luna possono alli viventi intervenire, per l'eccellentiss. Astrologo, Geomante, Chiromante, et Fisionomo M. Annibale Raimondo Veronese* (Venezia 1550) 104ᵛ.

84. Filippo Baldinucci, *Vocabolario toscano dell'arte del disegno* (1681), in *Opere*, 2: 60.

85. Baldinucci, *Vocabolario toscano dell'arte del disegno*, 2: 216.

86. Abbé Antoine-Joseph Pernety, *La Connaissance de l'Homme Moral par celle de l'Homme physique*, 2 vols. (Berlin 1776–1777) 1: 169. Cf. also a contemporary German testimony on the French usage of "air" and "mine": "L'air ist eine Gestalt des Gesichts, die man bey besonderen Fällen, um eine Passion an Tag zu legen, annimmt, und sind folglich so vielerley Arten desselben, als Affecten sind; man kann, zum Exempel, fröhliche, traurige, furchtsame, zornige, verliebte Gesichter machen, u.s.w. Solchergestalt differiren la mine, und l'air, daß die Mine beständig, l'air aber veränderlich" (Julius von Rohr, *Einleitung zur Ceremoniell-Wissenschafft der Privat-Personen* [Berlin 1728] 184).

87. Leopardi, *Zibaldone* [1666] 466.

88. "il nome di aria [. . .] si è dato a questa significazione generale di una fisionomia, appunto perché ella consistendo in sottilissimi rapporti colle qualità non materiali dell'uomo, è una cosa impossibile a determinarsi e quasi aerea" (Leopardi, *Zibaldone di pensieri* [1666–1667] 466).

89. Cf. for instance Scaligero, *Poetices*, 177: "quia conveniunt partes, corpus pulchrum est: igitur pulchritudinis causa convenientia." See here chapt. 1, above.

90. "questa significazione delle fisonomie, ch'è del tutto diversa dalla bellezza assoluta, e non è altro che un rapporto messo dalla natura fra l'interno e l'esterno, fra le abitudini ec. e la figura; questa significazione, dico, è una parte principalissima della bellezza, una delle capitali ragioni per cui questa fisonomia ci produce la sensazione del bello, e quella il contrario. Non è mai bella fisonomia veruna, che non significhi qualche cosa di piacevole [. . .]; ed è sempre brutta quella fisonomia che indica cose dispiacevoli, fosse anche regolarissima" (Leopardi, *Zibaldone* [1510–1511] 429).

91. "siccome l'interno degli uomini perde il suo stato naturale, e l'esterno più o meno lo conserva, perciò la significazione del viso è per lo più falsa; e noi sapendo ben questo allorchè vediamo un bel viso, e nondimeno sentendocene egualmente dilettati (e forse talvolta egualmente commossi), crediamo che questo effetto sia del tutto indipendente dalla significazione di quel viso, e derivi da una causa del tutto segregata ed astratta, che chiamiamo bellezza. E c'inganniamo interamente perchè l'effetto particolare della bellezza umana sull'uomo [. . .] deriva sempre essenzialmente dalla significazione ch'ella contiene, e ch'è del tutto indipendente dalla sfera del bello, e per niente astratta nè assoluta" (Leopardi, *Zibaldone* [1512] 429).

92. "la significazione della fisonomia nasce in gran parte dalle assuefazioni, cioè dal carattere, dalle passioni ec. ec. che l'individuo acquista appoco appoco, e che mettono in azione, e danno rappresentanza alla fisonomia" (Leopardi, *Zibaldone* [1905] 518).

93. "si va quasi distruggendo [. . .] la principal distinzione che la natura ha posto fra le cose animate e inanimate, fra la vita e la morte, cioè la facoltà del movimento" (Leopardi, *Zibaldone* [1607] 452).

94. From Latin "continentia."

95. Cf. Matthew Arnold's rhetorical question: "what is a countenance without its expression?" "The Study of Poetry," *The Portable Matthew Arnold*, ed. Lionel Trilling (New York: Viking 1949) 300.

96. I quote from the second edition: *A Pleasant History: Declaring the whole Art of Physiognomy, Orderly uttering all the speciall parts of Man from the Head to

the Foot (London 1613), f. 85r, 90r. On Hill cf. Francis R. Johnson, "Thomas Hill: An Elizabethan Huxley," *The Huntigton Library Quarterly* 7 (1944) 329–351, who lists also a 1556 work, "An Epitomie of the whole Arte of Phisiognomie," translated out of Cocles, which I have not seen. According to Dolores Beck Yonker, *The Face as an Element of Style: Physiognomical Theory in Eighteenth Century British Art* (Diss., Los Angeles: UCLA 1969) 48, it was "claimed by the transl. as the first work on the subject in English." Cf. also Nicholas Hilliard, *A Treatise concerning the Arte of Limning*, ed. R. K. R. Thornton and T. G. S. Cain [Manchester: Carcanet New Press 1981] 77: "How then can the curious drawer watch, and as it were catch these lovely graces, witty smilings, and those stolen glances which suddenly like lightning pass, and another countenance taketh place, except to behold and very well note and conceit to like?"

97. Joh. Alberti Bengelii *Gnomon Novi Testamenti* (1742; Tübingen 1855) 96; J. A. Bengel, *Gnomon of the New Testament*, vols. 5 (1859; Edinburgh 1860), ed. Andrew R. Fausset; trans. James Bandinel, 1: 313. To provide additional evidence from the lexicographical literature, the *Dictionary English-Latin and Latin-English* by Elisha Coles (1679) gives for "countenance" "vultus, aspectus, gestus," and as examples "A Letter will keep its Countenance, Epistola non erubescit" and "His countenance comes and goes, Non constat ei nec color nec vultus."

98. *Oxford English Dictionary*. More probably, it derives from Italian "baia," meaning "burla, beffa," according to the *Vocabolario della Crusca*, formed on the verb "baiare < abbaiare," imitating the barking of dogs, used in the locution " 'abbaiare dietro a uno' per canzonarlo" (Battaglia, *Grande Dizionario della Lingua Italiana*). Cf. also D. Enrico Zaccaria, *L'elemento germanico nella lingua italiana* (1901; Bologna: Forni 1986), *ad vocem* "beffa."

99. "das schnelle Maulaufsperren und -zuschließen (Bah)" (*Anthropologie in pragmatischer Hinsicht*, in *Gesammelte Schriften* [Berlin: Königlich Preußische Akademie der Wissenschaften 1907] 7: 301). The exact portraiture of the expression coordinated to the sound can be seen in a loose sheet, on which Lichtenberg drew a whole series of such sound-images: "ho ho/sni sni/bah bah/heng heng/heing heing" [fig. 15] (cf. *Georg Christoph Lichtenberg 1742–1799: Wagnis der Aufklärung* [München: Hanser 1992] 179).

100. *Plutarch's Moralia*, 7: 47.

101. Interpreting "out of countenance" (with John Wilkins) as "not knowing which way to look" (*An Alphabetical Dictionary*, bound with *An Essay towards a Real Character, and a Philosophical Language* [London 1668; rpt. Menston: Scolar Press 1968] *ad vocem* "countenance").

Chapter 4

1. Cf. Aristotle, *Probl.* XXVI.38 and *Meteor.* I.13; cf. also Bacon, *Historia ventorum* (1622) in his *Works*, ed. Spedding (1870; rpt. 1968) II.ii: 74 (English trans. by Francis Headlam, V.ii: 196: "the motion of winds is in most respects seen in the motions of water, as in a mirror"). In the microcosm of man, passions play the role of the winds: "as these purge and purifie the Air, so those cleanse and defecate the Blood, and suffer it not by stagnation to corrupt." Henry More, *An Account of Virtue*

(London 1690) 34. The "trace of a transient mood" (*vestigium transeuntis animantis*) pertains to the natural signs, according to Augustine, because it occurs independently from the conscious will (*vultus irati seu tristis affectionem animi significat, etiam nulla ejus voluntate qui aut iratus aut tristis est*). The human countenance marks thus the borderline between natural and conventional signs (Augustine, *De doctrina christiana* II.2, *PL* 34, c. 36).

2. "dicimus autem et vultum coeli et vultum maris, quia et mare saepe in varios motus ventorum flatibus mutatur et coeli vultus ex luce in tenebras et ex sereno in nubilum commutatur, sicut et hominum cum mentibus vultus." *Differentiarum, sive de proprietate sermonum, libri duo* (*PL* 83: c. 68).

3. "Vultuoso" has survived only in Spanish among the Romance languages. Still in Isidorus's sense, Ficino eloquently writes in his commentary *Sopra lo amore* that "our countenance takes on many colors [like the cloudy air, when the sun shines opposite to it, creates the rainbow] (*Il volto non altrimenti di varii colori si veste, che si faccia lo aere nebuloso, quando per aver il sole averso, crea lo arco baleno*)." The passage between squared brackets is only in the Italian version. Sears R. Jayne's translation trivializes matters beyond due measure: "our faces turn many colors like the rainbow when the sun shines opposite the misty air." *Marsilio Ficino's Commentary on Plato's Symposium* (Columbia: U of Missouri P 1944) 201.

4. Many illustrations can be found in Barbara Obrist, "Wind Diagrams and Medieval Cosmology," *Speculum* 72 (1997) 33–84.

5. Alberti, *On Painting* II.45. Praxiteles had been able, according to Pliny, to represent the "aurae velificantes" the garment of one of his statues (*N.H.*, XXXVI.29).

6. *Historia ventorum*, II:ii: "venti humanae genti alas addiderunt" (Engl. trans., V.ii: 139). The treatise was published in the third part of his *Instauratio Magna*, whose frontispiece proudly heralds two ships crossing at full-sails the once feared Hercules's columns.

7. Cf. esp. II.ii: 20 (Engl. trans. V.2: 145: "for the sake of clearness and to assist the memory, we give a new set of names to the winds according to their order and degrees, instead of using the old proper names"). I reproduce a diagram of the winds inspired by similar concerns from G. B. Della Porta's *De aeris transmutationibus*, Roma 1610, p. 31.

8. From Shelley's West Wind through Wordsworth's "corresponding" and Coleridge's "intellectual breeze" down to Mallarmé's *brise marine* and Valéry's *vent* tout court.

9. *Purg.* XI 94ff: "un fiato / di vento, ch'or vien quinci e or vien quindi,/e muta nome perché muta lato." Cf. Aristotle, *Meteor.* 1042b22, and Isidorus, *Etymologiae* XIII.xi: "Ventus [est] aer commotus et agitatus, et pro diversis partibus caeli nomina diversa sortitus." On the individuality of the winds in the Aristotelian meteorology as opposed to the Presocratics' view of the sea-of-air, cf. Victor Courant and Val L. Eichenlaub's "Introduction" to their edition of Theophrastus, *De ventis* (Notre Dame: Notre Dame UP 1975) XLI–XLII. There we find the following definition: "the breezes (*aurai*) arise when the moisture is condensed" (25).

10. *The Situations and Names of Winds* (*Ventorum Situs et Cognomina*) 973a1–973b25 in *Minor Works* (Cambridge, Mass.: Harvard UP 1936) 453–457.

11. In his treatise on Italian ortography against Trissino: Agnolo Firenzuola, *Discacciamento de le nvove lettere, invtilmente aggivnte ne la lingva toscana* (1524),

in *Trattati sull'ortografia del volgare 1524–1526*, ed. Brian Richardson (Exeter: U of Exeter 1984) 25. Jean de Meun already aimed at standardizing what he perceived as the extravagant richness of the Latin nomenclature of winds in his translation of Vegetius's *De re militari* (1284) (cf. Folena, " 'Volgarizzare' e 'tradurre'," 72); a task Bacon set out to fulfil with his *Historia*.

12. "καλοῦμεν [. . .] αὔρας τὰς ἐξ ὑγροῦ φερομένας ἐκπνοάς." [Aristotle], *On the Cosmos* 394b12–14; trans. D. J. Furley (Cambridge, Mass.: Harvard UP 1955) 367.

13. Certainly not strong enough to pose a threat. Cf. Hans Blumenberg, *Arbeit am Mythos* (Frankfurt/M.: Suhrkamp 1979) 49–50: "Die Welt mit Namen zu belegen, heißt, das Ungeteilte aufzuteilen und einzuteilen, das Ungriffige greifbar, obwohl noch nicht begreifbar zu machen. Auch Setzungen der Orientierung arbeiten elementaren Formen der Verwirrung, zumindest der Verlegenheit, im Grenzfall der Panik, entgegen. Bedingung dessen ist die Ausgrenzung von Richtungen und Gestalten aus dem Kontinuum des Vorgegebenen. Der Katalog der Winde, der günstigen und der ungünstigen [. . .] ist Kennzeichen einer Lebenswelt, in der Witterung Schicksal werden kann."

14. Cf. Ernst R. Curtius, *Europäische Literatur und Lateinisches Mittelalter* (Bern: Francke 1948) 105–108.

15. *The General History of the Air*, Designed and Begun by the Honble Robert Boyle Esq. (London 1692) 1.

16. Ennius still considers *aer* a foreign word (*Ann.* 148: "uento, quem perhibent Graium genus aera lingua"), though he starts to decline it as a Latin noun (*aerem* in *Epich. Var.* 56): cf. J. Marouzeau, *Quelques aspects de la formation du latin littéraire* (Paris: Klincksieck 1949) 129.

17. Pierre Chantraine, *Dictionnaire étymologique de la langue grecque*, *ad vocem* ἀήρ: "le mot signifie toujours chez Homére le brouillard et notamment la vapeur qui s'élève du sol et reste en suspension dans la partie la plus basse de l'atmosphère." Cf. also Paraskevaides, *The Use of Synonyms in Homeric Formulaic Diction,* 102–103.

18. Commenting on an extraordinarily "darke thick miste happening on the 27 of november 1674," Browne expands to comment upon the origin and nature of these phenomena, which "may well deserve the first place as being if not the first in nature, yet the first meteor mentioned in Scripture, and soone after the creation, for it is said Genesis 2. that God had not yet caused to raine upon the earth, but a mist went up from the earth and watered the whole face of the grounde, for it might take a longer time for the elevation of vapours sufficient to make a congregation of clowdes able to afford any store of showers and rayne in so early dayes of the world. Thick vapours not ascending high but hanging about the earth and covering the surface of it are commonly called mists; if they ascende high they are termed clowdes. They remaine upon the earth, till they either fall downe or are attenuated, rarified and scattered.[. . .] Mist[s] make an obscure air but they beget no darknesse, for the atomes and particules thereof admit the light, but if the matter therof be very thick, close, and condensed, the mist growes considerably obscure and like a clowde." "Upon the darke thick miste happening on the 27 of november 1674," *The Works of Sir Thomas Browne*, 5: 197.

19. In his first encounter with Rosenkranz and Guildenstern, *Hamlet* II. ii.

20. "The circumambient air doth make us all,/To be but one bare individual." Sir John Suckling, cited in Thomas Blount, *Glossographia: or a Dictionary, Interpret-*

ing all such Hard Words, Whether Hebrew, Greek, Latin, Italian, Spanish, French, Teutonick, Belgick, British or Saxon, as are now used in our refined English Tongue (London 1656), *ad vocem* "circumambient." I have not been able to locate the lines in any of Suckling's published works.

21. Leo Spitzer does not mention the English translation in his otherwise exhaustive essay. See also André Chastel, "*L'Aria*: Théorie du Milieu à la Renaissance" (1973), in *Fables, Formes, Figures*, 2 vols. (Paris: Flammarion 1978) 1: 393–405. "The circumambient air" of the Greeks, I suggest, is a synecdoche for "the circumambient world" Worsdworth refers to in *The Prelude* (1850) VIII.56. We can nowadays avoid such a periphrasis by using the term "environment," which was however not yet available to Wordsworth, at least in 1805, having been first introduced into the English language by Carlyle in 1827 (Spitzer, "Milieu and Ambiance," 232–233). The closest approximation to "environment" we have in ancient writing is the juxtaposition of *Airs, Waters, Places* in the title of a treatise that belongs to the Hippocratic corpus and is indeed devoted to what we would now qualify as environmental concerns.

22. "[. . .] which included all cloud, and rain, and dew, and darkness, and peace, and wrath of heaven." *The Queen of the Air: Being a Study of the Greek Myths of Cloud and Storm* (1869) in Ruskin's *Works*, ed. E. T. Cook and Alexander Wedderburn (London: Allen 1905) 19: 327.

23. Ruskin, *Works*, 19: 328.

24. Cf. the impassioned defence of the legitimacy of his argument in a letter written May 18, 1871: "The "Queen of the Air" was written to show, not what could be fancied, but was felt and meant, in the myth of Athena. Every British sailor knows that Neptune is the god of the sea. He does not know that Athena is the goddess of the air; I doubt if many of our school-boys know it—I doubt even if many of our school-masters know it, and I believe the evidence of it given in "The Queen of the Air" to be the first clear and connected appassionate proof of it which has yet been rendered by scientific mythology, properly so called." *Arrows of the Chace*, in *Works*, 34: 504.

25. Thomas Traherne, *Centuries, Poems, and Thanksgivings*, ed. H. M. Margoliouth, 2 vols. (Oxford: Clarendon 1958) 2: 152.

26. Cf. Hopkins's "The Blessed Virgin compared to the Air we Breathe," l. 10.

27. Aristotle, *Generation of Animals*; trans. Peck, 398–401.

28. W. H. S. Jones, *The Medical Writings of Anonymus Londinensis* (Cambridge: Cambridge UP 1947) 37. An Aristotelian origin of the comparaison has been suggested: cf. Jeanne Ducatillon, "Le traité des *vents* et la question hippocratique," *Formes de pensée dans la collection hippocratique: Actes du IV colloque international hippocratique (Lausanne, 21–26 septembre 1981)* (Genève: Droz 1983) 266–267; but a Stoic origin seems indeed more likely (as suggested by Walter Spoerri in the ensuing discussion [276]).

29. Cf. Rohde, *Psyche*, 548–550; Hans Diller, "Die philosophiegeschichtliche Stellung des Diogenes von Apollonia", *Hermes* 76 (1941): 359–381; Volker Langholf, *Medical Theories in Hippocrates: Early Texts and the 'Epidemics'* (Berlin: De Gruyter 1990) 244–245.

30. Hippocrates, *Breaths*; trans. W. H. S. Jones, 2: 231.

31. Which shows how the interest for questions we might now define of public health had become already acute by the second half of the seventeenth century in England.

32. John Evelyn, *Fumifugium: or, the Inconvenience of the Aer, and Smoake of London Dissipated*, London 1772, 1st ed. 1661 (rpt.: Oxford: Oxford UP 1930) 14.

33. *The Hous of Fame* II. 257–260, in *The Complete Works of Geoffrey Chaucer*, ed. Walter W. Skeat (Oxford: Clarendon 1926) 3: 23.

34. *The Hous of Fame* II. 323–325, *Complete Works*, 3: 25.

35. "Fugue for One Voice," *A Summoning of Stones* (New York: Mac Millan 1954) 8. Cf. "Songs for the Air or Several Attitudes About Breathing," in the same collection.

36. *Macbeth* I.vi.

37. *Hamlet* V.ii.

38. Evangelista Torricelli is the student of Galilei who invented the barometer and discovered the so-called "Torricellian vacuum." The discoveries of the atmospheric void and the atmospheric pressure thus do not belie each other, as Eugenio Montale suggests in his poem "Il vuoto," an elegy mourning the loss of the void: "È sparito anche il vuoto/dove un tempo si poteva rifugiarsi./Ora sappiamo che anche l'aria/è una materia che grava su di noi./Una materia immateriale, il peggio/che poteva toccarci." *Tutte le poesie* (Milano: Mondadori 1977) 607.

39. Tommaso Campanella, *Del senso delle cose e della magia* (Bari: Laterza 1925) 182.

40. I do not address the category of the aether in this context since it never straightforwardly crossed the path of physiognomy, although such doctrines as that of an aethereal body are clearly relevant to the discourse of physiognomy. In a sense, the aether is by definition so imperceptible to a bare eye that it becomes unusable as a physiognomical category. Air and aura are still perceivable even as they fade.

41. Albertus Magnus, *Metaphys.* XI.2.3 (cit. in Guido Cavalcanti, *Rime*, ed. Domenico De Robertis [Torino: Einaudi 1986] 17).

42. Cf. Enzo Volpini's article *ad vocem* "aria" in the *Enciclopedia dantesca* I: 365–368.

43. Cf. Verg., *Georg.* 4.417.

44. Blake's expression, quoted by Yeats in his essay "Blake's Illustrations."

45. *Le canzoni di Arnaut Daniel* X.43, critical ed. by Maurizio Perugi, 2 vols. (Milano: Ricciardi 1978) 2: 333.

46. Cf. Paolo Cherchi, "Gli 'adynata' dei trovatori," *Andrea Cappellano, i trovatori e altri temi romanzi* (Roma: Bulzoni 1979) 19–51; Barbara Spaggiari, "Cacciare la lepre col bue," *Annali della Scuola Normale Superiore di Pisa*, Classe di Lettere e Filosofia, Serie III, vol. 12:4 (1982) 1333–1403, who stresses the Plutarchean origin of the topos (*De tranq. an.* 471D); to Plutarch was even attributed an unparalleled collection of *adynata* (1344). The translation proposed by Perugi for "amassar" is "to embrace."

47. Trans. Robert M. Durling, *Petrarch's Lyric Poems: The Rime sparse and Other Lyrics* (Cambridge, Mass.: Harvard UP 1976), slightly modified.

48. Cf. Gianfranco Contini, "Préhistoire de l'*aura* de Pétrarque" (1955), reprinted in *Varianti e altra linguistica: Una raccolta di saggi (1938–1968)* (Torino: Einaudi 1970) 195.

49. Ed. Perugi, IX.1–8.

50. Trans. Mandelbaum, 2: 246, slightly modified.

51. Cf. the almost literal recovery of this motive in the final paragraphs of Castiglione's *Cortegiano*: "a soft breeze seemed to come that filled the air with a brisk coolness and began to awaken sweet concerts of joyous birds in the murmuring forests of the nearby hills (*parea che spirasse un' aura soave, che di mordente fresco empiendo l'aria, cominciava tra le mormoranti selve de' colli vicini a risvegliar dolci concenti dei vaghi augelli*)" (trans. Singleton).

52. *Anthology of the Provençal Trobadours*, ed. Raymond Thompson Hill and Thomas Goddard Bergin (New Haven: Yale UP 1941) 11.

53. Trans. Mandelbaum, slightly modified (Mandelbaum scales down the action of the breeze to a mere "striking," as Durling does in the sonnet by Petrarch I quote below).

54. Cf. also the usages of *ferire* to describe the action of the wind in Verg., *Aen.* 1.103, and Lucanus, *Phars.* 9.877 and 10.245: "zephyri aquas [. . .] Nili [. . .] feriunt." Cf. also Boccaccio, *Filocolo* II.25.

55. Trans. Durling, slightly modified, *Petrarch's Lyric Poems*, 342. Contini too hastily dismisses Dante's usage of *aura* in the *Commedia* as "toujours pourvue d'une tonalité modeste et immédiatament sémantique" (194–195). Petrarch also echoes Dante when he calls "l'aura soave che dal chiaro viso/move col suon de le parole accorte/per far dolce sereno ovunque spira," "quasi un spirto gentil di paradiso" (109). Cesare Segre limits himself to signal the presence of Dante's trope in Petrarch's sonnet in his "I sonetti dell'aura" (1983), *Notizie dalla crisi* (Torino: Einaudi 1993) 52. In the same volume cf. also "Le isotopie di Laura," 66–80. On the pervasiveness of Dante's presence in Petrarch, cf. Paolo Trovato, *Dante in Petrarca: Per un inventario dei dantismi nei "Rerum vulgarium fragmenta"* (Firenze: Olschki 1979).

56. Mario Equicola d'Alveto, *Di Natura d'Amore. Di nuovo ricorretto, e con somma diligenza riformato* (Venetia 1587) 53.

57. Cf. art. דמה in *Theologisches Wörterbuch zum Alten Testament* (Stuttgart: Kohlhammer 1990ff.) 7: 282–283. On the enduring echo of this theophany in later mystical poetry cf. Paolo Valesio, "'O entenebrata luce c'en me luce': la letteratura del silenzio," *Del silenzio*, ed. Giovannella Fusco Girard and Anna Maria Tango (Salerno: Ripostes 1992) 15–44, esp. 29–30.

58. This occurrence is supposedly dependent from the former (cf. Jörg Jeremias, *Theophanie: Die Schilderung einer alttestamentlichen Gattung* [Neukirchen-Vluyn: Neukirchener Verlag 1965] 112.) 112–115. This is in turn dependent from the theophany on Mount Sinai in *Ex.* 33, 18 ff.

59. Which was widespread among the peoples surrounding Israel, and deeply influenced the more current forms of biblical theophany (cf. on this Jeremias, *Theophanie*, 73–90). The subtlety of this breath of air is such that it has escaped even the notice of a fine enquirer like Corrado Bologna, who simply bypasses the word in his otherwise exhaustive *Flatus vocis: Metafisica e antropologia della voce* (Bologna: Mulino 1992).

60. *ōs en aurai leptēi*. Pseudo-Dionysius Areopagita, *De divinis nominibus* I.6 (596B), *Corpus dionysiacum* I, ed. Beate R. Suchla (Berlin: De Gruyter 1990) 119. See also IX.1 (909B).

61. The scholiast to the Pseudo-Dionysius stresses that God, by manifesting itself in the breeze (*pnoē*), is still electing to manifest itself in a body, although a body that cannot be contained in other bodies, is participated by everybody (*para pantōn metechetai*), and is in everybody (*PG*, vol. 2 of Dionysius's *Opera omnia*, c. 208). The aura is the material vehicle of the divine epiphany, although at its most refined state.

62. Spitzer's term to describe the consequences of linguistic secularization: cf. his essay "Dieu possible. Die Grammatikalisierung der nomina sacra," in *Stilstudien* (München: Hueber 1961) 2: 126–145. Luciano Rossi has pointed out the importance of Ovid's example in the history of the *aura*-motive up to Boccaccio and Petrarch in his article "Per la storia dell''Aura'," *Lettere Italiane* 42 (1990) 553–574. However, he does not pursue the broader implications of this *exemplum*, which is far more than just an illustration of "the pernicious effects of jealousy."

63. Ovid, *Metamorphoses* VII.821: "vocibus ambiguis deceptam praebuit aurem." I quote from the Loeb Classical Library edition of the works of Ovid, 6 vols., vol. 3: 1; trans. Frank J. Miller (Cambridge, Mass.: Harvard UP 1977).

64. Ovid, *Ars amatoria* III.698 (Loeb 1979) 2: 167.

65. *Metamorphoses* VII.822–823.

66. *Metamorphoses* VII.830.

67. *Ars amatoria* III.729.

68. *Ars amatoria* III.741.

69. *Metamorphoses* VII.857.

70. Xenophanes fr. B 32.

71. Cf. Max A. Warburg, *Zwei Fragen zum "Kratylos"*, Ph.D. Diss. (Berlin 1929) 70.

72. Cf. Freud's observations on "the usage of language" in neurosis and other "archaic modes of thought" in "Character and Anal Erotism" (1908), *Standard Edition* (1959) 9: 174.

73. *Metamorphoses* VII.760. Cephalus situates his own destiny in reference to that of his countryman.

74. *Metamorphoses* VII.712–713: "si mea provida mens est,/non habuisse voles."

75. This instance perfectly fits Saussure's posthumously published theory of the hypogram in ancient poetry (cf. Jean Starobinski, *Les mots sous les mots: Les anagrammes de Ferdinand de Saussure* [Paris: Gallimard 1971]).

76. If we grant Max Müller's derivation of Daphne from a Sanskrit word meaning "dawn," the name "aurora" also contains an allusion to the laurel tree. See his "Philosophy of Mythology" (1871), *Selected Essays on Language, Mythology and Religion*, 2 vols. (London 1881) 1: 607–608. Cf. the fine observations by Antoinette Novara, "Sur des noms propres devenus noms communs à la suite de métamorphoses: Remarques à partir du Chant X des *Métamorphoses* d'Ovide," *Sens et pouvoirs de la nomination dans les cultures hellénique et romaine: Colloque du séminaire d'Etudes des mentalités antiques 23–24 mai 1987* (Montpellier: Université Paul Valéry 1988) 129–135.

77. The easiest way to explain the inconsistency Philo remarks is, of course, by acknowledging that the two passages are indeed due to two different authors, J and P. See Harold Bloom's "Introduction" to *The Book of J*; trans. David Rosenberg (New York: Grove Weidenfeld 1990) 9–55, esp. 24–29.

78. *Legum allegoria* I.33; trans. Colson and Whitaker, 1: 169–170.

79. *Legum allegoria* I.42, 1: 173.

80. *Aura* does not surface at all in Hans Leisegang's extensive discussion of the concept *pneuma* in Philo, to which he devotes the entire vol. 1 of his *Der heilige Geist: Das Wesen und Werden der mystisch-intuitiven Erkenntnis in der Philosophie und Religion der Griechen* (Leipzig: Teubner 1919).

81. Alexander von Humboldt, *Versuch über die gereizte Muskel- und Nervenfaser nebst Vermuthungen über den chemischen Process des Lebens in der Thier- und Pflanzenwelt*, 2 vols. (Berlin 1797) 1: 78–79.

82. "Unter allen physikalischen Versuchen, welche ich je die Freude hatte, in Gegenwart anderer Naturforscher anzustellen, habe ich keinen gefunden, der wegen seiner unendlichen Feinheit so in Erstaunen setzt, als diese Belegung mit Hauche. Die Kette trockner Metalle, als Gold, Zink und Gold bringt keine Reizung hervor. Man hauche leise die eine untere, oder obere Fläche [. . .] man lasse, das gasförmige Wasser [. . .] diese eine Fläche überziehen, so wird der Muskel convulsivisch erschüttert [. . .] Man wische den Hauch mit einem wollenen Tuche ab, so verschwindet die Bewegung von neuem. Das Experiment sieht einem Zauber ähnlich, indem man bald —Leben einhaucht, bald den belebenden Odem zurücknimmt!" Humboldt, *Versuche*, 1: 78–79.

83. *Goethes Briefwechsel mit Wilhelm und Alexander von Humboldt*, ed. L. Geiger (Berlin: Bondy 1909) 293.

84. The word "rainbowy" is used by William Taylor in his *Historic Survey of German Literature* apropos of the scenery of Klopstock's poetry, which is always illuminated by "a misty glory, an intangible rainbowy lustre" (London 1830) 1: 292.

85. Cf. for instance Torricelli: "l'aure [. . .] quasi sempre sù la spiaggia marittima in tempo di state si sentono venir dalla marina" ("Del vento," *Lezioni accademiche*, in *Opere di Evangelista Torricelli* [Faenza: 1919] 2: 56).

86. Quoted in Franz Rosenzweig, "Unmittelbare Einwirkung der hebräischen Bibel auf Goethes Sprache," in M. Buber-F. Rosenzweig, *Die Schrift und ihre Verdeutschung* (Berlin: Schocken 1936) 130.

87. Cf. Harris Francis Fletcher, *Milton's Rabbinical Readings* (Urbana: U of Illinois P 1930) 119, 126–127.

88. In the sense now obsolete of "marked by genius," German *genial*.

89. Herder does not use *sausen* (cf. however: "das *Säuseln der Lüfte!*", *Älteste Urkunde des Menschengeschlechts*, in Johann Gottfried Herder, *Werke*, 10 vols. [Frankfurt/M.: Deutscher Klassiker Verlag 1993] 5: 203), but Rosenzweig seems to imply that by assonance Goethe might have derived it from *brausen*.

90. Rosenzweig, "Unmittelbare Wirkung," 133–134.

91. Buber-Rosenzweig, *Das Buch im Anfang*.

92. Valéry Larbaud, *Sous l'invocation de Saint Jérome* (Paris: Gallimard 1946) 176.

93. Like the graf of a silver rib in the body of language (Benjamin's best definition of his own practice of quotation, in *Einbahnstraße*, *Gesammelte Schriften*, IV·1: 131).

94. Hugo von Hofmannsthal, "Unsere Fremdwörter" (1914), *Gesammelte Werke*, ed. Herbert Steiner, *Prosa* III (Frankfurt/M.: Fischer 1952) 197.

95. "in der schöpferischen Laune des Augenblicks aus dem Gefüge einer fremden Sprache herausgerissen, nur für diesen einmaligen Gebrauch genau die vom Sprechenden gewollte Nuance des Ausdrucks." "Unsere Fremdwörter," 197. Fifteen years later Béla Balász will quote almost word by word this passage in a conversation with Benjamin (or is Benjamin's memory that translates Balász's speech in Hofmannsthal's terms?): "er brachte eine sehr richtige Beobachtung über das Fremdwort vor: daß es nämlich als solches immer aus seiner natürlichen Sprachbewegung herausgerissen und ein starres Gebilde sei" (*Gesammelte Schriften* VI: 418). "Ein starres Gebilde" is, in all likelihood, Benjamin's interpolation.

96. Goethe's interest in Humboldt's experiment seems to be chiefly stimulated by the thought of an analogy between the combined action of the breath and the hand of the experimenter, on the one hand, and the action of the atmospheric pressure, on the other, which by "rubbing" produces the colors of the soap-bubbles (*Goethes Briefwechsel*, 291).

97. Cf. Alberto Zamboni, *L'etimologia* (Bologna: Zanichelli 1976) 85–86.

98. "das Moment des Unauslöslichen, der untilgbaren Farben," Adorno, *Noten zur Literatur*, 26; trans. Shierry Weber Nicholson, "The Essay as Form," *Notes to Literature*, 2 vols. (New York: Columbia UP 1991) 1: 17.

99. "Über den Gebrauch von Fremdwörtern," *Noten zur Literatur*, 645.

100. See above, p. 26.

101. This might replace his formula, according to which "the *result* of the word is apprehended and retained, but the *schematismus* by which that result was ever reached is lost." T. de Quincey, "English Dictionaries," *Notes from the Pocket-Book of a Late Opium-Eater*, *Collected Writings*, ed. David Masson, 14 vols. (Edinburgh 1890) 10: 434.

102. Aristotle's term for foreign words in the *Poetics* XXI–XXII. See above, pp. 7–8.

103. Cf. Benjamin's definition of the "angelic tongue" in his essay on *Karl Kraus* as the tongue "in which all words, startled from the idyllic context of meaning, have become mottoes in the book of Creation" (*Reflections*, 269.) Here Benjamin is dependent, I believe, on Hermann Güntert's discussion of the *Glossolalie* as *Engelsprache* in his *Von der Sprache der Götter und Geister* (Halle: Niemeyer 1921) 23–31, a book Benjamin was well familiar with and quotes in the prologue to the *Ursprung des deutschen Trauerspiels*.

104. In these terms Alfred Polgar once characterized the peculiarity of Benjamin's style, "an amazing intuition," Benjamin himself confirms, "which, taking my style as its starting point, penetrates into the depth of my nature (*eine erstaunliche Intuition, die vom Stil ausgehend in meine Tiefe dringt*)" (*Gesammelte Schriften* VI: 418).

105. "On the Origin of Beauty: A Platonic Dialogue," *The Journal and Papers of Gerard Manley Hopkins* (London: Oxford UP 1966) 97–98.

106. Bacon, *Of Beauty*. "What we try to do in Aesthetics," according to Wittgenstein, "is to give *reasons*, e.g., for having this word rather than that in a particular place in a poem, or for having this musical phrase rather than that in a particular place in a piece of music." G. E. Moore, "Wittgenstein's Lectures in 1930–33," *Mind* 64 (1955) 19.

107. I quote from Edmund Jephcott's translation (*Reflections*, 94).

108. Grimm, *ad vocem* "Glück."

109. This is, for instance, the term Freud uses when discussing Dostoyevsky's epilepsy in "Dostojewski und die Vatertötung" (*Gesammelte Werke* [London: Imago 1948] 410).

110. Galen, *De locis affectis* III.ii, Kühn vol. 8, p. 194. Quoted in Owsei Tomkin, *The Falling Sickness: A History of Epilepsy from the Greeks to the Beginning of Modern Neurology* (Baltimore: Johns Hopkins UP 1945) 36.

111. Woolf, "On Being Ill," 11. Cf. chapt. 1, p. 14.

112. The title of Tomkin's book, *The Falling Sickness*, which is a literal translation of this new name of the illness, sounds almost as uncanny as Benjamin's translation of the name of its symptom.

113. "In den Vernichtunsnächten des letzten Krieges erschütterte den Gliederbau der Menschheit ein Gefühl, das dem Glück der Epileptiker gleichsah. Und die Revolten, die ihm folgten, waren der erste Versuch, den neuen Leib in ihre Gewalt zu bringen" (*Gesammelte Schriften* IV·1: 148).

114. Cf. Leo Spitzer, "Un hapax ne peut être expliqué par la linguistique seule: *espaciun* dans Gormont of Isembart," *Romanische Literaturstudien 1936–1956* (Tübingen: Niemeyer 1959) 28–32.

115. Paul de Man, "Conclusions: Walter Benjamin's 'The Task of the Translator'," *The Resistance to Theory* (Minneapolis: U of Minnesota P 1986) 92.

116. *Reflections*, 312.

117. De Man limits himself to quote only the first paragraph of the fragment in support of his reading of Benjamin's essay on "The Task of the Translator," according to which poetic language only originates "within the negative knowledge of its relation to the language of the sacred"—this being "a necessarily nihilistic moment that is necessary in any understanding of history" (*Resistance to Theory*, 92). But I do not think that the idea of happiness can be so easily dispensed with. In the paragraphs De Man does not quote Benjamin is clearly still paraphrasing the final chapter of Bloch's work, titled "Karl Marx, Death, and the Apokalypse," where he could find what he calls "a mystical conception of history." The very same words, "*restitutio in integrum* out of the labyrinth of the world," are indeed used by Bloch to indicate "the goal (*Ziel*) that matters most in the organization of life on earth," thus implying the necessity of a "downfall" of the labyrinthian physical world in which we are wandering—"this laborious, deathly-clear ruble heap of worlds which God shattered" (Ernst Bloch, *Geist der Utopie* [Berlin: Duncker & Humblot 1918] 442, 436; trans. E. B. Ashton, *Man on his Own* [New York: Herder & Herder 1970] 70, 64). As the thoroughly revised, 1923 edition of *Geist der Utopie* makes clear, Bloch was here alluding to the Kabbalistic doctrine of the breaking of the vessels.

118. *Reflections*, 313.

119. Etymologically, *eür*, from which *bon eür* and then *bonheur*, derives from *augurium*, but already Nicole Oresme seems to have been unaware of this relationship (cf. Alice Planche, "Eür, bon eür et bonheur dans un corpus lyrique de Moyen Âge tardif", in *L'idée du bonheur au Moyen Âge*, ed. Danielle Buschinger [Göppingen: Kümmerle 1990] 355). "La beauté n'est que la *promesse* du bonheur" is Stendhal's famous *bon mot* from *De l'amour*, on which cf. T. W. Adorno, *Ästetische Theorie* (Frankfurt a/M: Suhrkamp 1970) 461.

CHAPTER 5

1. Cf. David M. Greene, "The Identity of the Emblematic Nemesis," *Studies in the Renaissance* 10 (1963) 26.

2. Plato, *Leges* 717D3.

3. *Schadenfreude* is Herder's translation of Greek ἐπιχαρεικακία (in his essay on "Nemesis. Ein lehrendes Sinnbild," *Zerstreute Blätter*, Zweite Sammlung [1786], in *Sämmtliche Werke*, ed. Bernhard Suphan [Berlin 1888] 15: 398), which Leo Spitzer has shown to be the source of the German word ("Schadenfreude," *Essays in Historical Semantics*, 135–146).

4. *Eth. Eudemia* 1233b24–25; trans. H. Rackham (Cambridge, Mass.: Harvard UP 1935) 20: 351.

5. On the chthonic aspects of Nemesis see especially Károly Kerényi, "Die Geburt der Helena," *Mnemosyne* 7 (1939) 161–179.

6. Hesiod, *Theog.* vv. 223–224; *Op. et dies* vv. 199–201.

7. Cf. Greene, "The Identity of the Emblematic Nemesis," 30–31.

8. Greene, 32. Already in 1428, though, in response to a query from Ciriaco d'Ancona, Francesco Filelfo had been able to draw a clear distinction between Nemesis, "the goddess of indignation, whom your Greeks call *nemesis* (*indignationis dea, quam graeci tui* νέμεσιν *nemesin uocant*)," and Fortuna: see the letter in *Epistolarum familiarum libri* (Venetiis 1502) 3v; and cf. Michele Feo, "Ramnusia. Identità medievale di una divinità problematica," *Studi latini in ricordo di Rita Cappelletto* (Urbino: Quattro Venti 1996) 167.

9. Cf. Erwin Panofsky, *The Life and Art of Albrecht Dürer*, 2 vols. (Princeton: Princeton UP 1955) 1: 80–82; id., "'Virgo & Victrix': A Note on Dürer's *Nemesis*," *Prints*, ed. Carl Zigrosser (New York: Holt, Rinehart and Winston 1962) 13–38.

10. In his diary of the travel to the Netherlands: Dürer, *Schriftlicher Nachlass*, 1: 183.

11. I quote this and the following passage from Panofsky's translation in "'Virgo & Victrix'," 15–16.

12. Following her transferral to Rome, the divinity was closely associated with the Emperor: "The awesome power of Nemesis, in the just overthrow of those who merit destruction, i.e., the enemies of the state, thereby bringing victory and peace to the Empire, was, like so many other divine figures, successfully enlisted in the cause of the Roman state and its leader." Michael B. Hornum, *Nemesis, the Roman State, and the Games* (Leiden: Brill 1993) 40.

13. Franz Kafka, *Der Process*, chapt. "Advokat; Fabrikant; Maler." Who does not remember the equally unsettling epiphany of the Statue of Liberty in the first chapter of Kafka's American novel,—whose original title, *The Disappeared (Der Verschollene)*, has been re-established by the editors of the critical edition of his works—in which the goddess of freedom is described as holding a sword, instead of the torch?

14. Ausonius, *Epigram* 42; trans. White. I refer, for this and the other ancient texts I quote, to Hornum's convenient catalog of the literary evidence in *Nemesis, the Roman State, and the Games*, 127. The Greek original (anonymous, *Anth. Graeca* XVI.263) can be found there at p. 149.

15. Theaetetus Scholasticus, *Anth. Graeca* XVI.221; trans. Paton (Hornum, 140).

16. Straton, *Anth. Graeca* XII.193; trans. Paton (Hornum, 119–120).

17. Cf. the anonymous epigram in *Anth. Graeca* XVI.223; trans. Paton (Hornum, 148): "Nemesis warns us by her cubite-rule and bridle neither to do anything without measure (*ametron ti poiein*) nor to be unbridled in our speech (*achalina legein*)."

18. Angelo Poliziano, *Manto*, v. 25, in *Silvae*, ed. Francesco Bausi (Firenze: Olschki 1996) 9.

19. Pliny, *Nat. Hist.*; trans. Rackham, 1: 15.

20. Pliny, *Nat. Hist.* XXVIII.v.22; trans. Rackham, 8:17. Ausonius echoes Pliny's surprise (*Mos.* 379): "et Latiae Nemesis non cognita linguae."

21. Horace, *Epist.* 2.1.156. For a correction of this view from the historian's perspective, cf. Susan E. Alcock, *Graecia Capta: The Landscapes of Roman Greece* (Cambridge: Cambridge UP 1993).

22. III.156–58; trans. Robert Fitzgerald.

23. III.159; trans. Fitzgerald.

24. *The Cypria*, fragm. 8, in *The Homeric Hymns and Homerica*; trans. Hugh G. Evelyn White, slightly modified (Cambridge, Mass.: Harvard UP 1982) 499.

25. And to Homer: cf. *Iliad* XI.649 and XIII.122. On this point see Kerényi, "Die Geburt der Helena," 166. But cf. Jean-Claude Turpin, "L'expression ΑΙΔΩΣ ΚΑΙ ΝΕΜΕΣΙΣ et les 'actes de langage'," *Revue des Études Grecques* 93 (1980) 352–367.

26. Nonnos, *Dionysiaca* XLVIII. 382; trans. W. H. D. Rouse, 3 vols. (Cambridge, Mass.: Harvard UP 1940) 3: 453.

27. "I am ashamed to have the name of bride who once was vergin (*aideomai methepein meta parthenon ounoma nymphēs*)." V. 905, 3: 489.

28. Cf. v. 248: "a manlike maid she was, who knew nothing of Aphrodite (*apeirētē Aphroditēs*)" (3: 443).

29. Georg Zoega straightforwardly identifies Psyche's antagonist with Nemesis in his essay "Tyche und Nemesis": "jene Aphrodite, die nach der gemeinen Fabel gegen Eros und Psyche zürnte und nachmals von ihnen versöhnt wurde, Nemesis war." *Goerg Zoegas Abhandlungen*, ed. Friedrich Gottlieb Welcker (Göttingen 1817) 45.

30. Apuleius, *Metam.* IV.30. I quote Walter Pater's translation of "The Story of Cupid and Psyche" in *Marius the Epicurean: His Sensations and Ideas* (New York: Doubleday 1935) 44.

31. Ennodius, *Car.* I.iv.84: "Discant populi tunc crescere divam/Cum neglecta iacet." Quoted in C. S. Lewis, *The Allegory of Love* (Oxford: Oxford UP 1959) 78.

32. Pausanias, *Graeciae descriptio* XXXIV.1 (Hornum, 117).

33. Pliny, *Nat. Hist.* XXXVI.xvii.7–8 (Hornum, 108).

34. Pliny, *Nat. Hist.* XI.ciii.251 (Hornum, 107). That surface is the table, Herder writes, "where she marks down for herself, quietly but indelibly, all human thoughts and deeds." "Nemesis," 428. In the passage just preceding the quote Pliny writes that "memory is seated in the lobe of the ear, the place that we touch in calling a person to witness."

35. An epigrammatic definition of the goddess from the *Anthologia Graeca* VII.630 (the author Antiphilus of Byzantium, Hornum, 101).

36. In Callimachus's *Hymn to Demeter* 6.56 Eryssichton taunts Demeter, and "Nemesis recorded (*egrapsato*) his evil speech" (Hornum, 95).

37. Linné straightforwardly identifies Nemesis with the *Jus talionis*: "Nemesis Divina Jus Talionis est." Carl von Linné, *Nemesis Divina*, ed. Wolf Lepenies and Larf Gustafsson, based upon the first complete Swedish editions by Eris Malmeström and Telemak Fredbärj (1968) (Munich: Hanser 1981) 104.

38. Herder was aware of Linné's work, as yet unpublished, because of its mention in one of the latter's autobiographies (Herder, "Nemesis," 331).

39. "Ihr furchtbarer Name ist nur durch Mißverstand furchtbar geworden." Herder, 330–331.

40. "Eine Schwester der Schaam." Herder, 427.

41. Emerson speaks of a "beautiful Nemesis," an echo of Tibullus's *Nemesis formosa*, in his *Journals and Miscellaneous Notebooks*, ed. Ralph H. Orth and Alfred R. Ferguson (Cambridge, Mass.: Harvard UP 1971) 9: 453.

42. "Die ehemalige Venus also ist jetzt in eine tugendhafte, keusche Göttin verwandelt." Herder, 26.

43. "Jugendlich kommt sie vom Himmel, tritt vor den Priester und Weisen/ Unbekleidet, die Göttin; still blickt sein Auge zur Erde./Dann ergreift er das Rauchfaß und hüllt demütig verehrend/Sie in durchsichtigen Schleier, daß wir sie zu schauen ertragen." Goethe, *Gedichte*, 1: 206. It might be interesting to remember, as another instance of *fausse reconnaissance*, that the first publication of these lines occurred under the title "Die Wahrheit" (cf. Bernhard Suphan, "Aeltere Gestalten Goethe'scher Gedichten," *Goethe-Jahrbuch* 2 [1883] 115).

44. "Er muß sich selbst zügeln lernen, auch wenn Hoffnung seine Schritte beflügelt." Herder, 417.

45. Goethe is quoting with approval a remark by Tischbein: *Italienische Reise*, 29 December 1786. While in Rome, Goethe tried to secure for the prince Carl August of Weimar a copy, if not an original, of an ancient Nemesis. See the letter to the prince from Rome, 16 December 1786, in Goethe's *Weimarer-Ausgabe*, IV. Abtheilung: 8. Band (1890) 84–85, where Goethe justifies the modern practice of copying on the basis of the ancients' own example: "Der Bildhauer Trippel hat eine kleine Nemesis in Marmor nach einer größern im Museo gearbeitet und man kann sagen, sie ist beßer als das Original, welches deswegen nicht übertrieben ist, da viele mittelmäßige Künstler, ja Handwercker in Alten Zeiten nach guten Originalen kopirten, ja zuletzt Copie von Copie gemacht ward, so kann an einer Statue die Idee schön, Proportion und Ausführung aber schlecht seyn und ein neuerer Künstler kann ihr einen Theil der Vorzüge wiedergeben, die ihre ganz verlohrnen Originale hatten."

46. "Wie von unsichtbaren Geistern gepeitscht, gehen die Sonnenpferde der Zeit mit unsers Schicksals leichtem Wagen durch, und uns bleibt nichts, als mutig gefaßt die Zügel festzuhalten und bald rechts, bald links, vom Steine hier, vom Sturze da, die Räder wegzulenken. Wohin es geht, wer weiß es? Erinnert es sich doch kaum, woher es kam." Goethe, *The Autobiography of Johann Wolfgang von Goethe*; trans. John Oxenford, slightly modified, 2 vols. (Chicago: U of Chicago P 1974) 2: 437.

47. See the detailed critical remarks appended to Zoega's essay in Welcker's edition: "Anmerkungen zu Herders Abhandlung über die Nemesis im zweyten Theil seiner Zerstreuten Blätter" (Zoega, 60–75).

48. "Die Quelle jeglicher Gerechtigkeit, die Gesetzgeberin des Weltalls, die Mutter des Schicksals." Zoega, 72. It is evident from such a formulation that Schelling's own appraisal of the goddess (see below) is indebted to Zoega's analysis.

49. "Alle Götter sind Hieroglyphen, und ihre Namen bedeuten nach Stellung und Umgebung: demohngeachtet bleibt eine ursprüngliche und bleibende Bedeutung in der einzeln stehenden Hieroglyphe, jedoch auch diese durch Zeiten, Orte und Meynungen bestimmt." Zoega, 32.

50. Zoega, 54.

51. Zoega, 36.

52. Zoega, 40–41.

53. Cf. especially Emile Benveniste, *Noms d'agent et noms d'action en Indo-Européen* (Paris: Maisonneuve 1948) 79–80; and Emmanuel Laroche, *Histoire de la racine NEM- en Grec ancien (νέμω, νέμεσις, νόμος, νομίζω)* (Paris: Klincksieck 1949).

54. Zoega 41, 73. Zoega seems to consider the name Adrastea as a loan from the Oriental religions of Aegypt and Phoenicia, rather than a name that could be etymologised in Greek, as it was already by the ancients, to mean the "Inescapable."

55. Macrobius I.19 (Zoega, 39). Zoega has been first identified as a source by Karl Borinski, "Goethes 'Urworte. Orphisch'," *Philologus* 69 (1910) 1–9.

56. This is a question Borinski does not address in his otherwise exhaustive essay. According to Welcker, Zoega's text was written in July 1794: see his *Zoegas Leben. Sammlung seiner Briefe und Beutheilung seiner Werke*, 2 vols. (Stuttgart 1819) 2: 432.

57. Tibullus plays on what must have been already a well-established *topos* when he invokes Spes against Nemesis, the *senhal* of his reluctant beloved in the second book on his elegies: "Hope promises me that Nemesis shall be kind; but she says Nay. Ah me! worst not the goddess, cruel girl." Tib. II.vi.27–28. Trans. J. P. Postgate, *Catullus, Tibullus, and Pervigilium Veneris* (Cambridge, Mass.: Harvard UP 1976) 281.

58. Hornum, 140.

59. Hornum, 227–228.

60. Cf. above, p. 90, note 62.

61. Nos. 46 and 44 in the first (1531) edition.

62. Herder, "Nemesis," 428. Cf. the *Greek Anthology* IX.146 (Hornum, 146–147), an anonymous epigram describing the two goddesses in a similar *contraltare* (Italian has a specific word to refer to such an arrangement) position: "I, Eunus, have set up Elpis and Nemesis by the altar, the one in order that thou mayst hope, the other that you mayst get nothing." Cf. also F. H. Marshall, "Elpis-Nemesis," *The Journal of Hellenic Studies* 33 (1913) 84–86.

63. It is interesting to remember that Goethe chose to have a Nemesis engraved as the frontispiece to the seventh volume of his *Neue Schriften* [fig. 23], published in 1800: cf. the letter to his Berlin publisher Unger on April 2, 1800, in *Weimarer-Ausgabe* IV. Abtheilung: 15. Band (1894) 51–52, which expresses mixed feelings about the outcome: "Die Nemesis kam zur rechten Zeit an, ich glaube sie soll das Tittelkupfer des siebenten Bandes recht erwünscht zieren. Wäre man freylich beysammen und könnte unter der Arbeit sich, von der einen Seite über die Intention, von der andern über die Möglichkeit der Ausführung besprechen; so würde in einzelnen Theilen noch etwas vollkommneres geliefert werden können, doch bey einer kleinen Arbeit, die blos zur Zierde bestimmt ist, wird man es wohl nicht aufs schärfste nehmen."

64. I quote the translation by Christopher Middleton: "Yet the repulsive gate [i.e., of ΑΝΑΓΚΗ, Necessity] can be unbolted/ Within such bounds, their adamantine wall,/Though it may stand, that gate, like rock, for ever;/One being moves, unchecked, ethereal:/From heavy cloud, from fog, from squall of rain/She lifts us to herself, we're winged again,/You know her well, to nowhere she's confined—/A wingbeat—aeons vanish far behind." Goethe, *Selected Poems* (Boston: Suhrkamp 1983) 233.

65. The term *aeon* was used by Herder in his 1801 dramatic "Allegorie" *Aeon und Aeonis*, and defined as "ein Zeitlauf von vielen Jahren" (*Sämmtliche Werke*, ed. Suphan, 28: 247–263 [Berlin 1884] 247). Aeonis is the daughter of Aeon and Arete, whose name is changed by the end of the drama to that of Agape. The play was first published in the first issue of Herder's journal *Adrastea*, an enterprise that testifies to his continuous interest in the goddess. But his was not the only journal named after her in those years, in which the dramatic vicissitudes of world history had to remind everybody of her prerogatives: in 1814, Heinrich Luden, professor of history at the University of Jena, started in Weimar the publication of the political journal *Nemesis*, an initiative that Goethe disapproved (cf. Luden's report of their November 1813 conversation and Goethe's questioning of the timeliness of the new publication in *Gespräche*, 3: 97–108; cf. also Goethe's remark to Riemer on January 5, 1814, "apropos of the journal *Nemesis* and of the unease that somebody had expressed about the title:" "Germans are ruminating animals [*Die Deutschen sind wiederkäuende Thiere*].") But nobody was certainly more mindful than Hölderlin of the name whose oblivion he countered through the creation of one of his most daring composite words: *Vergangengöttliches* (*Germanien*, v. 100). In *Hyperion* it is Diotima's task to admonish his lover by first evoking her, but the entire novel is under the spell of the goddess who gives the name to the "brotherhood of Nemesis" (*Bund der Nemesis*).

66. *Orat.* 64.8.1–2. Trans. H. Lamar Crosby, *Dio Chrysostom*, 5 vols. (Cambridge, Mass.: Harvard UP 1951) 5: 51.

67. Zoega, 55.

68. Macrobius, *Saturnalia* I.xxii: "Nemesis, quae contra superbiam colitur, quid aliud est quam solis potestas, cujus ista natura est ut fulgentia obscuret et conspectui auferat, quaeque sunt in obscuro illuminet offeratque conspectui?" Hornum, 133.

69. "Nemesis ist nichts anderes als die Macht eben jenes höchsten, alles in Bewegung bringenden Weltgesetztes, das nicht will, daß irgend etwas verborgen bleibe, das alles Verborgen zum Hervortreten antreibt und gleichsam moralisch zwingt sich zu zeigen." Schelling, *Philosophie der Mythologie*, in *Sämmtliche Werke* (Stuttgart 1857) II.2: 146–147.

70. Macrobius's source has been in turn identified as a (lost) treatise by Porphiry *On the Divine Names*: cf. Pierre Courcelle, *Les lettres grecques en Occident. De Macrobe a Cassiodore* (Paris: Boccard 1948) 17–20.

71. "Nihil enim est opertum, quod non revelabitur: et occultum, quod non scietur." Matt. 10, 26.

72. From the *Tuba mirum*: "Whatever is hidden shall be made known,/nothing shall remain unpunished." On the likely connections of the hymn with the preaching of Gioacchino da Fiore cf. Kees Vellekoop, *Dies Ira, Dies Illa: Studien zur Frühgeschichte einer Sequenz* (Bilthoven: Creyghton 1978).

73. Schelling, 142.

74. Schelling, 145; a view that is shared by contemporary scholars: cf. Benveniste and Laroche, quoted above, no. 53.

75. "Das Unmittelbare," as opposed to "die strenge Mittelbarkeit" that is the law (*das Gesez*), in the annotation to his Pindar's translations (*Grosse-Stuttgarter-Ausgabe* 5:285).

76. Schelling, 144–145.

77. As such, Nemesis is certainly not compatible with chance, she is "die Macht, die das bloß Zufällige nicht duldet" (Schelling, 152).

78. *Pyth.* X.45; *Olymp.* VIII.86.

79. Schelling, 649.

80. Whereas, according to Schelling, the order ought to be reversed: "Nicht er hat die Mythologie erzeugt, sondern er selbst ist das Erzeugnis der Mythologie, und zwar jener letzten Krisis," namely, the oblivion of monotheism (Schelling, 648).

81. "Die homerische Welt schließt schweigend ein Mysterium in sich, und ist über einem Mysterium, über einem Abgrund gleichsam errichtet, den sie wie mit Blumen bedeckt" (Schelling, 649).

82. "Homeros, d.h. der homerische Polytheismus beruht gerade nur auf diesem Vergessenseyn des Mystischen" (Schelling, 648).

83. Cf. Kerényi, 164–165.

84. Schelling's definition "throws a quite new light on the concept of the *Unheimlich*, for which we were certainly not prepared." *The "Uncanny"*; trans. James Strachey, *Standard Edition*, 17: 225.

85. Freud quotes Schelling from Sanders's *Wörterbuch der deutschen Sprache*, which elides (by chance?) the clause "in der Latenz." In its entirety the text reads: "unheimlich nennt man alles, was im Geheimnis, im Verborgenen, in der Latenz bleiben sollte und hervorgetreten ist." Schelling, *Philosophie der Mythologie*, 649.

86. Freud, *"Uncanny,"* 236. Cf. Heine, *Die Götter im Exil* in *Historisch-kritische Gesamtausgabe*, ed. Manfred Widfuhr, vol. 9 (Düsseldorf: Hoffmann und Campe 1987) 125.

87. Cf. J.-A. Hild, *Étude sur les démons dans la littérature et la religion des Grecs* (Paris 1881) 2: "Chez Homère [. . .] daimon n'apparaît qu'à des rares intervalles, avec un sens spécial et mystérieux qui embarrasse le traducteur moderne, et même pour le poète antique ne semble pas exempt d'obscurité;" cit. in Gilbert François, *Le polythéisme et l'emploi au singulier des mots ΘΕΟΣ, ΔΑΙΜΩΝ* (Paris: Belles Lettres 1957) 53.

88. Apuleius, *De deo Socratis* XV.150–151. Augustine might be mindful of Apuleius's tentativeness in this passage when he explains the absence of the word *daimōn* from the title of the treatise as due to shame, because "non est Socratis amicitia daemonis gratulanda," *Civ. Dei* VIII.14. On the history of demonology in the pagan world cf. the "Introductory Notes" to J. Den Boeft, *Calcidius on Demons (Commentarius Ch. 127–136)* (Leiden: Brill 1977) 1–7. The part of the body consecrated to the Genius is the forehead, according to Servius: "frontem Genio (consecratam esse), unde uenerantes deum tangimus frontem " (Serv., Aen. 3, 607), cit. in George Dumézil, *Religion roman archaïque* (Paris: Payot 1966) 365. *Genius*, according to the definition of Censorinus, is "deus, cuius in tutela ut quisque natus est uiuit" (*De die natali* 3, 1) 364.

89. Orig. *In Matth.*, 13, 27–28. Cf. Robert Schilling, "Genius et Ange," first in Pauly-Wissowa, s.v. *Genius* (1976), now in *Rites, Cultes, Dieux de Rome* (Paris: Klincksieck 1979) 415–444: 432.

90. Leopardi's Tasso will thus be visited by a *genio familiare*. In his autobiography Goethe expressed the hope that the word, "which, though apparently foreign, really belongs to every people" (*Autobiography*, 2:405) may survive in spite of his contemporaries' abuse. In spite of the still ongoing abuse, it may have truly reached its nadir in the headline, calling a horse a genius, which convinces Musil's "man without qualities" to take a year of vacation from life (*Der Mann ohne Eigenschaften*, chapt. 13).

91. Aristophanes, *Frogs*, 1182–1188; trans. Benjamin Bickley Rogers (Cambridge, Mass.: Harvard UP 1930) 406–409. Cf. Marcel Detienne, *La notion de Daïmôn dans le pythagorisme ancien* (Paris: Belles Lettres 1963) 66–67.

92. *Tim.* 90C. The pun is, according to Detienne, of Pythagorean origin (Detienne 64). According to Simmias, disciple of Philolaos, the true name of the soul is *daimōn* (Plut., *De genio Socr.* 591D–E). What Plato calls *daimōn* in the *Timaeus* is, more specifically, the most refined part of the soul: the *nous*. The link between the archaic view and Plato's psychologization is Empedocles's view of the demon as "the occult self" that persists through successive reincarnations, whose function is "to be the carrier of man's potential divinity and actual guilt" (Dodds, *Greeks and the Irrational*, 153, 210, 213). According to Dodds, in Plato the demon becomes "a sort of lofty spirit-guide, or Freudian Super-ego" (Dodds, 42), whereas, if at all, it would have rather to be identified with the Id.

93. Such a development reflects the loss of a difference that is still evident in the formulaic epithet for heroes, *daimōni isos*, which is probably pre-Homeric: Diomedes, when compared to a *daimōn*, is threatening even to a god such as Apollo. François sees in such a usage "l'écho d'une conception primitive:" "daimon désigne vraisemblablement un type de puissance surnaturelle dont Homère n'a pas harmonisé la conception avec celle des dieux olympiens." François, *Le polythéisme*, 328–329.

94. Pierre Chantraine declares the derivation as a popular etymology and "peu vraisemblable," but "satisfaisante pour l'esprit." "Le divin et les dieux chez Homère," in *Entretiens sur l'Antiquité classique*, I (Geneva: Vandœuvres 1952) 45–94: 81. Wilamowitz adopts it and translates *daimōn* as *Zuteiler* (Ulrich von Wilamowitz-Möllendorff, *Der Glaube der Hellenen* 2 vols. [Berlin: Weidemann 1931] 1: 369). Heidegger adopts another etymology in his reading of the word (from *daiō*, "to present oneself in the sense of pointing and showing," *Parmenides*; trans. André Schuwer and Richard Rojcewicz [Bloomington: Indiana UP 1992] 102), but nevertheless concludes that we translate the demonic as "the uncanny (*das Unheimliche*)" (Heidegger 101).

95. A relationship that Creuzer's definition implicitly brings to the fore: "to the Greek, the demon was the superior and hidden power that directed his destiny independently of any participation on his side (*Aux yeux du Grec, le Démon était la puissance supérieure et cachée qui dirigeait sa destiné indépendamment de tout concours de sa part*)," G. F. Creuzer, *Religions de l'Antiquité*, ed. and trans. J. D. Guigniaut (Paris 1838) III.1: 4. Cf. also the relationship between *daimōn* and revenge (*daimones alastores*) discussed in Detienne 48–50 and 87–90.

96. Walter Burkert, *Greek Religion*; trans. John Raffan (Cambridge, Mass.: Harvard UP 1985) 180.

97. Schelling, 650 (emphasis mine).

98. Cf. *Od.* 14.443: *daimonie xeinōn*. It is a proverbial expression that "the gods do liken themselves to strangers from abroad" (*Od.* 17.485). In Homer *daimonios* is only used in the vocative case. Cf. Elisabeth Brunius-Nilsson, *Daimonie: An Inquiry into a Mode of Apostrophe in Old Greek Literature* (Uppsala: Almqvist 1955). The meaning of the epithet is correctly understood by Baudelaire in the first of his *Petits poèmes en prose*, "L'étranger," though he splits it in two: *énigmatique* and *extraordinaire*.

99. Goethe, *Autobiography*, 2: 423, slightly modified.

100. Goethe, 2: 425, also modified.

101. "Träger des Grauens ist der Name für jedes religiöse Volk" (M. A. Warburg, *Zwei Fragen zum "Kratylos,"* 72).

102. *Apology* 26B (one of Meletus's charges, according to Socrates).

103. "Man hat schon vor Alters gesagt: die Grammatik räche sich grausam an ihren Verächtern, Du spricht es in Deinem letzten Briefe durch das Wort *nemesisch* gar vortrefflich aus." *Briefwechsel zwischen Goethe und Zelter in den Jahren 1799 bis 1832*, 23 February 1832, *Sämtliche Werke: Münchner Ausgabe*, vol. 20.2 (Munich: Hanser 1998) 1618. The motto is a quote from Luther, which Goethe in turn almost certainly knew from Herder, who had used it against the pedantic classicist Klotz in the third of his *Kritische Wälder* (1769), *Sämmtliche Werke*, 3: 472. Klotz's treatise "De verecundia Virgilii" was one of the targets of Herder's satyre in the second "Wäldchen," under the title "Ueber die Schaamhaftigkeit Virgils" (3: 272–320).

104. In the line of a poem addressed to him, Herder apostrophized Goethe as "Thou, who from gods (*Götter*) art discended, or Goths (*Gothen*), or from origin filthy (*Koth*)," where the last term is obviously a charitable euphemism of the translator (Goethe 2: 14).

105. Goethe, 2: 14.

106. Wilamowitz, *Glaube der Hellenen*, 1: 362.

107. Dodds, *Greeks and the Irrational*, 13.

108. In commenting upon Mallarmé's comparison of the current usage of language to the exchange of a coin, whose sides are both effaced, and which is passed from hand to hand "en silence," Lacan subtly observes that "cette métaphore suffit à nous rappeler que la parole, même à l'extrême de son usure, garde sa valeur de tessère" (*Écrits* [Paris: Seuil 1966] 251).

109. It is interesting to notice a series of peculiarities in this *incipit*: Freud makes a point to remark that he is reversing the order of the actual inquiry in the course of his exposition, what came last in the process of discovery is here presented first: "my investigation was actually begun by collecting a number of individual cases, and was only later confirmed by an examination of linguistic usage." Moreover, Freud progresses from what is less known to what is more known to us, without justifying this departure from the Aristotelian-Hegelian method of exposition he, too, usually follows. Furthermore, he acknowledges that the lexicographical inquiry, whose results he presents, is not his own work, but Theodor Reik's, as if to disclaim any responsability for the conclusions he has to draw from the material his assistant had assembled. I suggest that all these features betray a certain anxiety on Freud's part, and that his desire to dismiss the search as useless too hurriedly, as it were, should be taken as a

token of repression, according to his own teaching. The question of the relationship to the German language was, of course, a *vexata quaestio* for the German-Jewish intellectuals, who were well aware of the nationalistic aspects of the linguistic debate in Germany, especially as far as the relationship to foreign loans is concerned. Hofmannsthal's essay is also meant to address this issue, which was particularly pressing in the years around World War I. Leo Spitzer defended the borrowing of foreign words against the linguistic xenophobia of writers such as Houston Stewart Chamberlain (cf. his *Anti-Chamberlain: Betrachtungen eines Linguisten über Houston Stewart Chamberlains "Kriegsaufsätze" und die Sprachbewertung im allgemeinen* [Leipzig: Reisland 1918]). The same dilemma resurfaced in Germany after the Second World War, when Adorno had to advocate once again the liberating power of the usage of foreign words.

110. Freud, *Gesammelte Werke* XII: 232; *Standard Edition* XVII: 221.

111. Ibid.

112. See Laplanche and Pontalis, *Language of Psychoanalysis*, for a list of translations, such as French *inquiétante étrangeté*, a solution that cannot but remind us of Amyot's compromise in translating *dysōpia*.

113. *The "Uncanny"*; trans. James Strachey, *Standard Edition*, 17: 225.

114. Freud, 17: 243. It is also interesting to notice that in the first edition the name "Schleiermacher" had replaced "Schelling" in this second occurrence, another paramnesia not to be underestimated.

115. E. Jones, "Fear, Guilt and Hate" (1929), *Papers on Psycho-Analysis* (5th ed., London: Baillière, Tindall, and Cox 1950) 312.

116. Though he gives two reasons for introducing the term: "In the first place I find it necessary to insist on the absoluteness of the thing feared, and this thing is something even wider and more complete than castration, if we use this word in its proper sense [. . .] the ultimate danger with which we are here concerned is to all possible forms of sexuality [. . .] It means total annihilation of the capacity for sexual gratification, direct or indirect [. . .] In the second place it is intended to represent an intellectual description on our part of a state of affairs that originally has no ideational counterpart whatever in the child's mind, consciously or unconsciously [. . .] I refer not merely to the birth situation itself, about which so much is still doubtful, but to many months afterwards when we can observe a state that may be called pre-ideational primal anxiety (*Urangst*). It is only later, when the situation is becoming externalized and the anxiety is created by the ego as a 'signal' (Freud) for warning purposes, that we can speak of ideational fear, one which then usually has a specific reference." Jones, 312–313.

117. In "The Early Development of Female Sexuality" (1927) Jones writes: "in both sexes castration is only a *partial* threat, however important a one, against sexual capacity and enjoyment as a whole. For the main blow of total extinction we might do well to use a separate term, such as the Greek word 'aphanisis'. If we pursue to its roots the fundamental fear which lies at the basis of all neuroses we are driven, in my opinion, to the conclusion that what it really signifies is this aphanisis, the total, and of course permanent, extinction of the capacity (including opportunity) for sexual enjoyment. [. . .] The male dread of being castrated may or may not have a precise female counterpart, but what is more important is to realise that this dread is only a

special case and that both sexes ultimately dread exactly the same thing, aphanisis." Jones, *Papers on Psycho-Analysis*, 440.

118. 259E, 2: 425–427. See p. 20 above. In Aristophanes, the noun refers to the material obliteration of the written text of a legal action: *aphanisis tēs dikēs* (*Nub.* 764), while in Tucidides the corresponding verb *aphanizō* also means "to obliterate" writing (VI.54).

119. Differently from what is implied by the translation, the verb points toward a complete erasure of expression, though voluntary, rather than a disfiguration: Liddell-Scott-Jones glosses the passage as referring to "artificial disfigurement."

120. Jacques Lacan, *Le Séminaire livre XI: Les quatres concepts fondamentaux de la psychanalyse* (Paris: Seuil 1973) 189.

121. Lacan, *Le Séminaire livre XI*, 199, 201.

122. Plutarch, *De latenter vivendo* 1130 E; trans. Einarson and De Lacy, "Is "Live unknown" a wise precept?" *Moralia*, 14: 341.

123. Léon Wurmser, *The Mask of Shame* (Baltimore: Johns Hopkins UP 1981) 84.

124. Wurmser, *The Mask of Shame*, 92.

125. Rilke, *Die Aufzeichnungen des Malte Laurids Brigge*, 6: 938; trans. Mitchell, *Notebooks*, 251, slightly modified.

126. Rilke, 6: 940; trans. Mitchell, 253.

127. Hornum, 131. Goethe dedicated a sonnet to the goddess (1808), in which Nemesis chases after the despiser of Amor, mentioned by its Latin name: see it in *Gedichte*, 1: 299–300.

128. *Inf.* V.103; trans. Robert Pinsky, *The Inferno of Dante* (New York: Farrar, Straus and Giroux 1994) 41. Alciati's emblem 111, whose subject is the triumph of *Anterōs*, the love of virtue, over Cupid, and whose author is declared to be Nemesis herself [fig. 24], implies already a moralization of the goddess, further removed from Dante's still genuine understanding of her power.

129. Roland Barthes, *Fragments d'un discours amoreux* (Paris: Seuil 1977) 129, 130 (trans. Richard Howard, *A Lover's Discourse: Fragments* [New York: Farrar, Straus and Giroux 1978] 112, 113). Howard further complicates matters by replacing "fading" in the original with "fade-out." Barthes defines the *fading* as a "painful ordeal in which the loved being appears to withdraw from all contact (*épreuve douloureuse selon laquelle l'etre aime semble se retirer de tout contact*)."

130. "Physiognomik der Stimme" is the title of the review Adorno devoted in 1957 to the German version of the first edition of Paul Moses's book (cf. note 132), now reprinted in his *Gesammelte Schriften* (Frankfurt/M.: Suhrkamp 1986) 20: 510–514.

131. Friedrich Arnold Klockenbring, "Von den verschiedenen Tönen der Aussprache des Worts, Ich," *Aufsätze, verschiednen Inhalts*, 2 vols. (Hannover 1787) 1: 160.

132. Paul J. Moses, *The Voice of Neurosis* (1954; New York: Grune and Stratton 1971) 37.

133. Byron's lines are probably meant as a rebuttal of Blake's optimistic identification of "The Divine Image" with "the human form divine." Blake himself had anticipated such a criticism in an early Song of Experience, where he defines "Terror, the Human Form Divine" (*The Complete Prose and Poetry of William Blake*,

newly revised edition, ed. David V. Erdman, commentary by Harold Bloom [New York: Doubleday 1988] 32).

134. Conrad, *Heart of Darkness, Complete Works*, 16: 149.

135. Suida (Hornum 141), *Anth. Graeca* VII.630 (Hornum 101).

136. Frg. 6 Diels.

137. Adrien Baillet, *Vie de Monsieur Descartes* (1691; Paris: Table ronde 1946) 282.

138. In a letter to Georg Brandes, dated Nizza, December 2, 1887: "eine Philosophie wie die meine, ist wie ein Grab—man lebt nicht mehr mit. *Bene vixit qui bene latuit*—so steht auf dem Grabstein des Descartes. Eine Grabschrift, kein Zweifel!"

139. Plutarch, *De latenter vivendo* 1128C; trans. De Lacy, 14: 325.

140. 1129F, 14: 335. Cf. the semi-equivalence (Detienne 130) between *genesis* and *daimōn* implicit in a passage of *Gen. Socr.* 585F.

141. Leibniz's definition of the atom; quoted by Max Scheler, in "Zur Rehabilitierung der Tugend: Die Demut; Die Ehrfurcht," *Vom Umsturz der Werte: Abhandlungen und Aufsätze* (1915; Bern: Francke 1972) 19.

142. Wurmser, *Mask of Shame*, 264.

143. Cf. Melvin R. Lansky and Andrew P. Morrison, "The Legacy of Freud's Writings on Shame," in *The Widening Scope of Shame*, ed. Lansky and Morrison (Hillsdale: Analytic Press 1997) 3–40.

144. Hesiod, *Op.et dies* 317.

145. Plutarch, 15: 115, 117.

146. Cf. Norbert Elias, *Über den Prozeß der Zivilisation*, vol. 1, published first in 1936; 2nd ed., 2 vols. (Bern: Francke 1969) and Hans Peter Duerr's five-volume rejoinder *Der Mythos vom Zivilisationsprozeß* (Frankfurt/M.: Suhrkamp 1988–2002).

147. "Verecundia ist die zentrale Erscheinungsform unseres Seelenlebens." Eugen Rosenstock(-Huessy), *Angewandte Seelenkunde. Eine programmatische Übersetzung* (Darmstadt: Roetherverlag 1924) 48. Rosenstock, Rosenzweig's friend and interlocutor, does not italicize the word in his text.

148. *Della Fisonomia di tutto il corpo humano del S.Gio.Batta.Della Porta Acc. Linceo Libri Quattro* (Roma 1637) 155: "Che li sfacciati divengano vergognosi."

149. Antonin Artaud, "Le visage humain" (1947), in *Du Visage*, ed. Marie-José Baudinet and Christian Schlatter (Lille: Presses Universitaires de Lille 1982) 9; trans. Clayton Eshleman with Bernard Bador, "The Human Face," in *Watch-fiends & Rack Screams: Works from the Final Period*, (Boston: Exact Change 1995) 277.

150. "die Schützgöttin der weiblichen Tugend." *Versuch einer Charakteristik des weiblichen Geschlechts*, 3 vols. (Hannover 1797–1799) 1: 186.

151. Spinoza writes that "pudor, quamvis non sit virtus, bonus tamen est, quatenus indicat, homini, qui pudore suffunditur, cupiditatem inesse honeste vivendi, sicut dolor, qui eatenus bonus dicitur, quatenus indicat, partem laesam nondum esse putrefactam" (*Eth.*, P. IV, Prop. LVIII, Scholium).

152. George Eliot's unfaithful but appropriate translation for *putrefactam* in Spinoza's Latin: cf. Eliot's unpublished translation of the *Ethics*, ed. Thomas Deegan (Salzburg: Universität Salzburg 1981) 196.

153. Mann, *Zauberberg*, 1: 43. The English translation by H. T. Lowe-Porter: "that grave, beautiful name of chastity," *The Magic Mountain* (New York: Knopf 1985) 39, here as elsewhere, is not trustworthy.

154. Conrad, *The Secret Agent: A Simple Tale*, 171.

155. "This dying virtue, this surviving shame/Whose crime will bear an ever-during blame." *Rape of Lucrece*, vv. 223–224.

156. Pindar, *Nem.* 11.45–46: δέδεται γὰρ ἀναιδεῖ ἐλπίδι γυῖα. Wilamowitz rightly observes that the qualifier here conveys much more than its later meaning "shameless" (*unverschämt*) may suggest (*Glaube*, 1: 356).

157. "Die Erlösung der Schönheit von der Sünde." Kerényi, "Die Geburt der Helena," 177.

158. *Iliad* III.383–420.

159. *Pandora*, v. 346: "Unmöglich's zu versprechen ziemt mir wohl." What Epimetheus is asking for himself to his own daughter is precisely "der Liebe Glück," the return of Pandora (v. 345).

160. "Sie sitzt und hiflos erhebt sie die Arme nach einer Frucht, die ihr unerreichbar bleibt. Dennoch ist sie geflügelt. Nichts ist wahrer." Benjamin, *Einbahnstraße*, *Gesammelte Schriften* IV·1: 125; *One-Way Street*; trans. Jephcott, *Selected Writings*, 1: 471. Benjamin's text follows almost word by word Leopoldo Cicognara's description of the relief in his *Storia della scultura dal suo risorgimento in Italia fino al secolo di Canova*, 7 vols. (2nd ed., Prato 1823–1825) 3: 399–400, which he could have found partially translated in French in a volume of engravings published in Florence in 1823: [Giovanni Paolo Lasinio] *Le tre porte del Battistero di San Giovanni di Firenze*, 8: " M.ʳ le Comte Cicognara dit que le Sculpteur voulant représenter l'Espérance s'est figuré un objet quelconque, une chose idéale à laquelle tendent les voeux pour l'obtenir. Il a placé vers elle une figure assise qui exprime toute la force du désir, qui incline le corps, étend les bras, leve les yeux et souhaite avec passion l'objet de son amour. Il semble qu'elle touche presque à son bout [here a section of the Italian text is missing, which makes reference to the wings of the figure: "col moto delle ali anche sembra che aggiugner voglia al movimento di tutta la persona."] Comment pourroit on mieux représenter l'Espérance?" [fig. 26].

Index

Abrams, M. H., 140n80
Addison, Joseph, 3, 70, 73, 124n13
Adorno, Theodor Wiesengrund, 6, 95,
  127n45, 179n98, 180n119, 189n109,
  190n130
adynaton, 82, 175n46
Aelian, 162n18
aeon, 185n65
aēr. See air
Aeschylus, 111
aether, 160n133, 175n40
Aetius, 30
Agamben, Giorgio, 41, 151n40
Agorakritos, 103
aidōs, 62, 97, 101–102, 118. *See also*
  shame
air, 11–12, 52, 54, 64–70 passim, 72–
  74, 82–85, 158n117, 159n119,
  164nn36–37, 164n39, 164n41,
  166n47, 170n86, 170n88, 173nn16–
  18, 175n40; and features, 70,
  164n38; in music, 87, 93, 95
Alberti, Leon Battista, 10, 19, 25, 27–
  28, 33, 50, 56, 77, 137n58, 141n89,
  158n115
Albertus Magnus, 85, 175n41
Alciati, Andrea, 190n128
Alcock, Susan E., 182n21

alētheia, 116
Ambrose, Saint, 46, 153n66
Amyot, Jacques, 61, 164n36, 189n112
anagnōrisis, 4
analogia, 24, 26, 140n81
anamnēsis, 9
anatomy, 14, 19, 56
Anaxagoras, 16–17
Anaximenes, 84
Anonymus Londinensis, 83
antonomasia, 31, 66, 82
aphanisis, 12, 20, 114–116, 189n117,
  190n118
Apollodorus, 155n86
apotheosis, 47, 123n6
Apuleius, 110, 186n88
aria. See air
Aristarch, 18
Aristides Quintilianus, 150n29
Ariston of Ceos, 62
Aristophanes, 111, 190n118
Aristotle, 4, 7–8, 18, 20, 31, 38, 42,
  46, 62, 66, 79, 83, 97, 119, 141n91,
  145n136, 146n143, 156n101, 165n45,
  171n1, 172n9, 179n102
Aristoxenus, 32
Arnold, Matthew, 170n95
Artaud, Antonin, 119, 191n149